Advances in
COMPUTERS
VOLUME 73

Advances in
COMPUTERS

Emerging Technologies

EDITED BY

MARVIN V. ZELKOWITZ

Department of Computer Science
University of Maryland
College Park, Maryland

VOLUME 73

ELSEVIER

AMSTERDAM • BOSTON • HEIDELBERG • LONDON • NEW YORK • OXFORD
PARIS • SAN DIEGO • SAN FRANCISCO • SINGAPORE • SYDNEY • TOKYO
Academic Press is an imprint of Elsevier

ACADEMIC
PRESS

Academic Press is an imprint of Elsevier
84 Theobald's Road, London WC1X 8RR, UK
Radarweg 29, PO Box 211, 1000 AE Amsterdam, The Netherlands
30 Corporate Drive, Suite 400, Burlington, MA 01803, USA
525 B Street, Suite 1900, San Diego, CA 92101-4495, USA

First edition 2008

ISBN: 978-0-12-374425-8

ISSN: 0065-2458

For information on all Academic Press publications
visit our website at elsevierdirect.com

Printed and bound in the United Kingdom

Contents

History of Computers, Electronic Commerce and Agile Methods

David F. Rico, Hasan H. Sayani and Ralph F. Field

Testing with Software Designs

Alireza Mahdian and Anneliese A. Andrews

Balancing Transparency, Efficiency and Security in Pervasive Systems

Mark Wenstrom, Eloisa Bentivegna and Ali R. Hurson

Computing with RFID: Drivers, Technology and Implications

George Roussos

Medical Robotics and Computer-Integrated Interventional Medicine

Russell H. Taylor and Peter Kazanzides

CONTENTS

Contributors

Anneliese Amschler Andrews is Professor and Chair of the Department of Computer Science at the University of Denver. Her current research interests include software testing, software design, software maintenance and empirical software engineering. She has published over 150 research papers in refereed software engineering journals and conferences. She serves in the editorial board of five software engineering journals and in many program committees for conferences. She received a PhD degree in Computer Science from Duke University.

Eloisa Bentivegna is a Ph.D. student in physics (with a minor in high-performance computing) at the Pennsylvania State University, State College. She received an M.Sc. in Theoretical Physics summa cum laude from University of Catania, Italy, in 2002. Her research areas include numerical relativity and cosmology. She was appointed as a John Archibald Wheeler Fellow in 2006 and has been an Eberly College of Science Duncan Fellow since 2003.

Ralph F. Field has 25 years of experience in education and program management. He has been working for the University of Maryland University College's Graduate School of Management and Technology since 1995, where he is an associate professor at both the masters and doctoral level. At University College, he is the Program Director for Not-for-Profit Management, Naval Operations and National Security, Army Sustaining Base Management, Joint Military Strategy, Planning and Decision Making and Air and Space Strategic Studies. He has performed extensive field work in Botswana, Africa, as a Peace Corp. volunteer. He holds a PhD in Development Sociology.

A. R. Hurson is the department chair and professor of the computer science department at the university of Missouri-Rolla. Before his current appointment, he was a Computer Science and Engineering professor at The Pennsylvania State University. His research for the past 25 years has been directed towards the design and analysis of general as well as special-purpose computer architectures. His research has been supported by NSF, NCR Corp., DARPA, IBM, Lockheed Martin, ONR and Penn State University. He has published over 250 technical papers in areas including database systems, multi-databases, global information-sharing processing, application of mobile-agent technology and object-oriented databases, Mobile-computing environment and computer architecture parallel and distributed processing. He is the co-author of the *IEEE Tutorials on Parallel Architectures for Database Systems, Multi-database Systems: An Advanced Solution for Global Information Sharing, Parallel Architectures for Data/Knowledge Base Systems, and Scheduling and Load Balancing in Parallel and Distributed Systems.*

He served as a member of the IEEE Computer Society Press Editorial Board, an IEEE distinguished speaker and editor of IEEE transactions on computers and IEEE/ACM Computer Sciences Accreditation Board. Currently, he is serving as an ACM lecturer, editor of Journal of Pervasive and Mobile Computing and editor of The CSI Journal of Computer Science and Engineering.

Peter Kazanzides received the B.Sc., M.Sc. and Ph.D. degrees in electrical engineering from Brown University in 1983, 1985 and 1988, respectively. His dissertation focused on force control and multi-processor systems for robotics. He began work on surgical robotics in March 1989 as a postdoctoral researcher at the IBM T.J. Watson Research Center with Dr. Russell Taylor. Dr. Kazanzides co-founded Integrated Surgical Systems (ISS) in November 1990 to commercialize the robotic hip replacement research performed at IBM and the University of California, Davis. As Director of Robotics and Software, he was responsible for the design, implementation, validation and support of the ROBODOC® hardware and software. Dr. Kazanzides joined the Engineering Research Center for Computer-Integrated Surgical Systems and Technology (CISST ERC) at Johns Hopkins University in December 2002. He currently holds an appointment as an Assistant Research Professor of Computer Science at Johns Hopkins University.

Alireza Mahdian is currently a second-year PhD student at the Department of Computer Science at the University of Denver. His current research interests include software testing, software design and software maintenance. He received his BS degree in Computer Engineering from Sharif University of Technology, Iran.

David F. Rico has been a systems engineer in support of the NSA, NRO, NASA, DARPA, DISA, SPAWAR, USAF, NAVAIR, CECOM and MICOM for 20+ years. He worked on NASA's $20 billion space station in the 1980s, he worked for a

$40 billion Japanese corporation in Tokyo in the early 1990s and he worked on U.S. Navy fighters such as the F-18, F-14 and many others. He's been an international keynote speaker, has published numerous articles and has contributed to five books on computer science. He holds a bachelor's degree in computer science, a master's degree in software engineering and a doctoral degree in information technology.

George Roussos holds a first degree in Pure Mathematics from the University of Athens, an MSc in Numerical Analysis and Computing from the University of Manchester and a PhD in Distributed Scientific Computing from Imperial College, where his studies were supported by a Marie Curie Fellowship. He is currently a Senior Lecturer at the School of Computer Science and Information Systems, Birkbeck College, University of London, where he leads the pervasive computing lab. Before joining Birkbeck College as a lecturer, he worked as the Research and Development Manager for a multi-national information technology corporation in Athens, Greece, where he was responsible for the strategic development of new IT products in the areas of knowledge management and mobile internet; as an Internet security officer for the Ministry of Defence, Athens, where he designed the Hellenic armed forces Internet exchange and domain name systems; and as a research fellow for Imperial College, London, where he conducted research in distributed systems. He is currently investigating the effects of social activity on system architectures and exploring mechanisms to support navigation and findability. He is a member of the ACM, SIGMOBILE, the IEEE, the IEEE Communications and the IEEE Computer Society.

Hasan H. Sayani's interests lie in information systems, development of information systems, life cycle methods and tools and semantic database management systems. He has taught at the University of Maryland – College Park in the Information Systems Management program. He also co-founded a commercial organization which built systems for various commercial and governmental organizations. He has participated in various professional (e.g., IEEE, ACM, CASE) and standardization organizations (e.g., ANS, CODASYL, DoD, CALS). He holds a BSE, MSE and PhD from the University of Michigan.

Russell H. Taylor received his Ph.D. in Computer Science from Stanford in 1976. He joined IBM Research in 1976, where he developed the AML robot language and managed the Automation Technology Department and (later) the Computer-Assisted Surgery Group before moving in 1995 to Johns Hopkins, where he is a Professor of Computer Science, with joint appointments in Mechanical Engineering, Radiology and Surgery, and is Director of the NSF Engineering Research Center for Computer-Integrated Surgical Systems and Technology. He is the author of more than 200 refereed publications, a Fellow of the IEEE and AIMB and a recipient of the Maurice Müller award for excellence in computer-assisted orthopaedic surgery.

Mark Wenstrom received his B.S. degree (summa cum laude) in computer science and engineering from Bucknell University, Lewisburg, in 2005. He received his M.E. degree in computer science and engineering from The Pennsylvania State University, State College, on May 2007.

Preface

This is volume 73 of the **Advances in Computers**. This series, which began publication in 1960, is the oldest continuously published anthology that chronicles the ever changing information technology field. In these volumes, we publish from 5 to 7 chapters that cover the latest changes to the design, development, use and implications of computer technology on society today. In this current volume, subtitled 'Emerging Technologies', we discuss several new advances in computer software generation as well as describe new applications of those computers.

In the first chapter, 'History of computers, electronic commerce and agile methods', D. F. Rico, H. H. Sayani and R. F. Field give an overview of various software development technologies that have been applied during the past 40 years, with the goal of improving the software development process. This includes various methods such as structured development methods, reviews, object-oriented methods and rapid development technologies. In this latter category, they spend the last third of their chapter reviewing the current development and interest in agile methods as a means to rapidly produce effective programs.

Anneliese Andrews and Alireza Mahdian in Chapter 2, 'Testing with software designs', explore implications of UML as an emerging design notation for software. As they state in their chapter 'Originally, designs in UML have been used to test implementations against their design artifacts, but there are also testing techniques that test the design artifacts directly'. They discuss techniques where designs in UML can be used to test the underlying implementation.

Chapter 3 'Balancing transparency, efficiency and security in pervasive systems', by Mark Wenstrom, Eloisa Bentivegna and Ali Hurson deal with the emerging concept of pervasive computing and its impact on resource management and security. The basic goals of pervasive computing are that computer technology is seamlessly available whenever and wherever the user is situated. But this goes against the security goals of isolating users from potentially malicious attacks by unauthorized individuals. Similarly, resources are not uniformly distributed throughout an environment, although computing resources are expected to be available when needed. In this chapter, the authors discuss how this goal of transparency of computers affects efficiency of the system as well as security concerns.

RFID, or Radio Frequency Identification, is coming to a store near you. This is the technology that cheaply tags products with unique identifiers that only need to pass near a reading device rather than specifically being read by a scanner. With this technology, products can be easily traced through the supply chain from the manufacturer to the user. George Roussos in Chapter 4, 'Computing with RFID: drivers, technology and implications', discusses this technology, how supply chains work in industry, and briefly gives an overview of the basic technology of its operation.

In addition to changes to your local supermarket described in the preceding chapter, robotic research will have an important impact on other aspects of everyday life. One area of growing use of robot control is in medicine. In the final chapter, Dr. Russell Taylor and Dr. Peter Kazanzides discuss the use of robot technology in medicine, specifically Computer-Integrated Interventional Medicine (CIIM), where robotic control takes over some or all of the aspects of surgery.

I hope you found this volume to be interesting. I am always looking for new and different chapters and volume themes to use for future volumes. If you know of a topic that has not been covered recently or are interested in writing such a chapter, please let me know. I am also always looking for qualified authors. If interested, I can be contacted at mvz@cs.umd.edu. I hope you like these volumes and I look forward to producing the next one in this long-running series.

Marvin Zelkowitz
University of Maryland
College Park, Maryland

History of Computers, Electronic Commerce and Agile Methods

DAVID F. RICO

HASAN H. SAYANI

Graduate School of Management and Technology
University of Maryland University College

RALPH F. FIELD

Graduate School of Management and Technology
University of Maryland University College

Abstract

The purpose of this chapter is to present a literature review relevant to a study of using agile methods to manage the development of Internet websites and their subsequent quality. This chapter places website quality within the context of the $2.4 trillion U.S. electronic commerce industry. Thus, this chapter provides a history of electronic computers, electronic commerce, software methods, software quality metrics, agile methods and studies on agile methods. None of these histories are without controversy. For instance, some scholars begin the study of the electronic computer by mentioning the emergence of the Sumerian text, Hammurabi code or the abacus. We, however, will align our history with the emergence of the modern electronic computer at the beginning of World War II. The history of electronic commerce also has poorly defined beginnings. Some studies of electronic commerce begin with the widespread use of the Internet in the early 1990s. However, electronic commerce cannot be appreciated without establishing a deeper context. Few scholarly studies, if any, have been performed on agile methods, which is the basic purpose of this literature review. That is, to establish the context to conduct scholarly research within the fields of agile methods and electronic commerce.

ADVANCES IN COMPUTERS, VOL. 73
ISSN: 0065-2458/DOI: 10.1016/S0065-2458(08)00401-4

1

1. Introduction

Agile methods are an approach for managing the development of new products based on principles of flexible manufacturing and lean development. The use of agile methods for Internet software was a reaction to the emergence of traditional software development methods, which were too cumbersome, expensive, rigid and fraught with failure. Downsizing was the norm and traditional methods were being used by large corporations in decline, rather than by young, energetic firms on the rise. Millions of websites were created overnight by anyone with a computer and a modicum of curiosity. Agile methods marked the end of traditional methods in the minds of their creators.

Traditional methods for managing software development were created when the first commercial computers began emerging in the 1950s. Scientists and engineers began creating increasingly more powerful and complex computer systems, and inordinately complex computer programs beyond the comprehension of a single human. These early computer programs had millions of components to perform the simplest of operations, giving rise to traditional methods. The rise of traditional methods is also linked to the debut of the commercial software industry in the 1960s. Traditional methods consisted of formal project plans, well-documented customer requirements, detailed engineering processes, hundreds of documents and rigorous testing.

Agile methods emerged with a focus on iterative development, customer feedback, well-structured teams and flexibility. Internet technologies such as HTML and Java were powerful new prototyping languages, enabling smaller teams to build bigger software products in less time. Because they could be built faster, customers could begin to see finished software sooner and provide earlier feedback, and developers could rapidly refine their software. This gave rise to closed-loop, circular, highly recursive and tightly knit processes for rapidly creating Internet software, leading to improvements in website quality for electronic commerce.

2. History of Computers and Software

2.1 Electronic Computers

Electronic computers are simply machines that perform useful functions such as mathematical calculations or inputting, processing and outputting data and information in meaningful forms [1]. As shown in Figure 1, modern electronic computers are characterized by four major generations: first-generation vacuum tube computers from 1940 to 1950, second-generation transistorized computers from 1950 to 1964, third-generation integrated circuit computers from 1964 to 1980 and fourth-generation microprocessor computers from 1980 to the present [1]. First-generation or vacuum tube computers consisted of the electronic numerical integrator and calculator or ENIAC; electronic discrete variable computer or EDVAC; universal automatic computer or UNIVAC; and Mark I, II and III computers [1]. Second-generation or transistorized computers consisted of Philco's TRANSAC S-1000, Control Data Corporation's 3600 and International Business Machine's 7090 [1]. Third-generation or integrated-circuit-based computers consisted of International Business Machine's System/360, Radio Corporation of America's Spectra 70 and

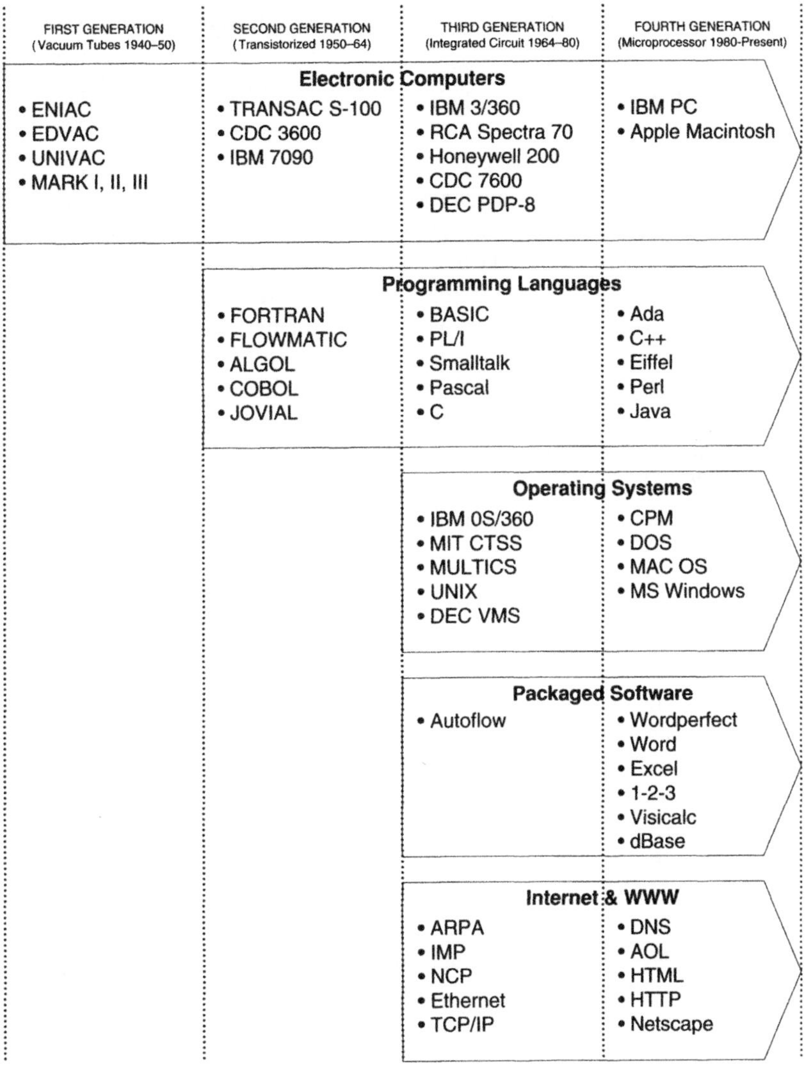

FIRST GENERATION (Vacuum Tubes 1940–50)	SECOND GENERATION (Transistorized 1950–64)	THIRD GENERATION (Integrated Circuit 1964–80)	FOURTH GENERATION (Microprocessor 1980-Present)
Electronic Computers			
• ENIAC • EDVAC • UNIVAC • MARK I, II, III	• TRANSAC S-100 • CDC 3600 • IBM 7090	• IBM 3/360 • RCA Spectra 70 • Honeywell 200 • CDC 7600 • DEC PDP-8	• IBM PC • Apple Macintosh
Programming Languages			
	• FORTRAN • FLOWMATIC • ALGOL • COBOL • JOVIAL	• BASIC • PL/I • Smalltalk • Pascal • C	• Ada • C++ • Eiffel • Perl • Java
Operating Systems			
		• IBM OS/360 • MIT CTSS • MULTICS • UNIX • DEC VMS	• CPM • DOS • MAC OS • MS Windows
Packaged Software			
		• Autoflow	• Wordperfect • Word • Excel • 1-2-3 • Visicalc • dBase
Internet & WWW			
		• ARPA • IMP • NCP • Ethernet • TCP/IP	• DNS • AOL • HTML • HTTP • Netscape

FIG. 1. Timeline and history of computers and software.

Honeywell's 200 [1]. Late third-generation computers included Cray's CDC 7600 as well as Digital Equipment Corporation's PDP-8, VAX 11/750 and VAX 11/780 [4]. Fourth-generation or microprocessor-based computers included the International Business Machine's Personal Computer or PC and Apple's Macintosh [3].

2.2   Programming Languages

Programming languages are defined as 'any of various languages for expressing a set of detailed instructions for a digital computer' [5]. By 1972, there were 170 programming languages in the U.S. alone [6] and today there are over 8500 programming languages worldwide [7]. First-generation or vacuum tube computers did not have any programming languages [6]. Second-generation or transistorized computers were characterized by an explosion of programming languages, the most notable of which included formula translation or FORTRAN, flowchart automatic translator or FLOWMATIC, algorithmic language or ALGOL, common business-oriented language or COBOL, Jules own version of the international algorithmic language or JOVIAL and the list processing language or LISP [6]. Third-generation or integrated-circuit-based computers likewise experienced a rapid increase in programming languages, the most notable of which were the beginner's all-purpose symbolic instructional code or BASIC, programming language one or PL/1, Smalltalk, Pascal and C [8]. Fourth-generation or microprocessor-based computers continued the trend of introducing new programming languages, such as Ada, C++, Eiffel, Perl, Java and C#.

2.3   Operating Systems

Operating systems are simply a layer of software between the computer hardware and end-user applications used for controlling hardware peripherals such as keyboards, displays and printers [2]. First-generation or vacuum tube computers did not have any operating systems and 'all programming was done in absolute machine language, often by wiring up plugboards' [3]. Second-generation or transistorized computers did not have any operating systems per se, but were programmed in assembly languages and even using the early computer programming language called formula translation or FORTRAN [3]. Third-generation or integrated-circuit-based computers consisted of the first formalized multi-programming operating systems and performed useful functions such as spooling and timesharing [3]. Examples of third-generation operating systems included IBM's Operating System/360, the Massachusetts Institute of Technology's compatible time-sharing system or CTSS, the multiplexed information and computing service or MULTICS, the uniplexed information and computer system or UNICS, which became UNIX, and Digital Equipment Corporation's virtual memory system or VMS [3]. Fourth-generation or microprocessor-based computers consisted of the control program for microcomputers or CPM, disk operating system or DOS, Apple's Macintosh operating system or MAC OS and Microsoft's Windows [3].

2.4   Packaged Software

Software is defined as 'instructions required to operate programmable computers, first introduced commercially during the 1950s' [9]. The international software industry grew slowly in revenues for commercially shrink-wrapped software from about zero in 1964, to $2 billion per year in 1979, and $50 billion by 1990 [10]. It is important to note that the custom, non-commercially available software industry was already gaining billions of dollars in revenue by 1964 [10]. First-generation or vacuum tube computers, much like programming languages and operating systems, did not have any software and 'all programming was done in absolute machine language' [3]. Second-generation or transistorized computers were characterized by bundled software, e.g., software shipped free with custom computer systems, and customized software such as International Business Machine's SABRE airline reservation system and the RAND Corporation's SAGE air defense system [10]. Third-generation or integrated-circuit-based computers saw the first commercialized shrink-wrapped software such as Applied Data Research's Autoflow flowcharting software [12] and the total annual sales for commercial software were only $70 million in 1970 compared with over $1 billion for custom software [10]. In part due to the U.S. Justice Department's anti-trust lawsuit against IBM around 1969, commercial software applications reached over 175 packages for the insurance industry in 1972 and an estimated $2 billion in annual sales by 1980 [10]. Fourth-generation or microprocessor-based computers represented the golden age of shrink-wrapped computer software and were characterized by Microsoft's Word and Excel, WordPerfect's word processor, Lotus' 1-2-3 and Visicorp's Visicalc spreadsheets, and Ashton Tate's dBase database [13]. By 1990, there were over 20 000 commercial shrink-wrapped software packages in the market [14]. And, the international software industry grew to more than $90 billion for pre-packaged software and $330 billion for all software-related products and services by 2002 [15] and is projected to reach $10.7 billion for the software as a service or SAAS market by 2009 [16].

2.5   Internet and WWW

The Internet is defined as a network of millions of computers, a network of networks, or an internetwork [17]. First-generation or vacuum tube computers were not known to have been networked. Late second-generation or transistorized computers gave rise to the Internet as it is known today [18]. Second-generation computers of the 1960s gave rise to packet switching theory, the first networked computers, the U.S. military's advanced research project's agency or ARPA and the first interface message processor or IMP [18], [19]. An MIT researcher published the first paper on packet switching theory, devising what was known as the 'Galactic Network' [18], [19]. This same researcher was appointed head of ARPA's Behavioural Sciences and Command and Control Programs [18]. The ARPANET was developed to see if machines could be networked and many machines, such as second-generation IBM 7090s were on the early ARPANET, even as third-generation computers began to emerge [18]. Third-generation or integrated-circuit-based computers took the early networking concepts devised during the second generation and formalized them into the ARPANET and Internet concepts as they are known today [18] All this came together when the Bolt, Beranek and Newman (BBN) Corporation installed the first IMP at UCLA in 1969 and the first host computer was connected [18]. The Network Working Group (NWG) completed the initial ARPANET host-to-host protocol called the Network Control Protocol (NCP) in 1970, network users began developing applications from 1971 to 1972 and the first public demonstration of the ARPANET took place in 1972 [18]. In summary, late third-generation computers of the 1970s gave rise to the network control protocol or NCP, email, open architecture networking, ethernet, transmission control protocol, Internet protocol and one of the first bulletin boards by Compuserve [18]. Late third-generation computers gave rise to the hyper text markup language or HTML. Tim Berners-Lee, a British physicist working for the Conseil Européen pour la Recherche Nucléaire or CERN (European Organization for Nuclear Research) created an early HTML prototype called Enquire, which ran on a Norwegian minicomputer called the Norsk Data Machine running the NORD Time Sharing System or NORD-TSS. HTML was first proposed to the Internet Engineering Task Force (IETF) in 1991 to 1993 and became an official standard in the 1995 to 1996 timeframe. Early fourth-generation or microprocessor-based computers gave rise to the domain name system or DNS and Prodigy and AOL were created [18]. Using middle fourth-generation computers, Tim Berners-Lee of CERN had created the first web server on a NeXTcube running the NeXTstep operating system, which was a UNIX variant, and is credited with the creation of the hyper text transfer protocol or HTTP in 1989. Middle fourth-generation computers of the Internet era were adapted to the formalized IETF HTML and HTTP standards and gave rise to Mosaic and Netscape, which caused the number of computers on the Internet to reach one million by 1992 and 110 million by 2001 [19]. Using ideas from Tim Berners-Lee, Marc Andreessen and Eric Bina, students at the National Centre for Supercomputing Applications (NCSA) at University of Illinois at Urbana-Champaign created the first popular WWW browser called Mosaic in 1992. Marc Andreessen formed Mosaic Communications Corporation to commercialize his WWW browser, which was renamed Netscape to deconflict with the NCSA's intellectual property claims. Netscape is credited with popularizing the WWW and Internet as it is known today.

3. History of Electronic Commerce

3.1 Electronic Commerce

The purpose of this section is to give a brief overview of the history of electronic commerce. Though electronic commerce seemed to enter into mainstream public consciousness in the 1990s, the electronic commerce industry is as old as the computer and software industries themselves. This section attempts to give readers a small appreciation of the earliest beginnings of the electronic commerce industry, as we believe electronic commerce is the key for the convergence of electronic computers, operating systems, programming languages, packaged software and the Internet and WWW (e.g., Apple iPhones). From a simple perspective, electronic commerce is defined as sharing of business information, maintaining business relationships or conducting business transactions using the Internet. However, there are at least four comprehensive definitions of electronic commerce [20]:

1. Communications perspective. Electronic commerce is the delivery of information, products, services or payments via telephones or computer networks.
2. Business process perspective. Electronic commerce is the application of technology to the automation of business transactions and workflows.
3. Service perspective. Electronic commerce is a tool that helps firms, consumers and managers cut service costs, improve quality and speed delivery.
4. Online perspective. Electronic commerce provides the capability of buying and selling products and information on the Internet and other online services.

Electronic commerce is one of the most misunderstood information technologies [20]. For instance, there is a tendency to categorize electronic commerce in terms of two or three major types, such as electronic retailing or online shopping [21]. However, as shown in Figure 2, electronic commerce is as old as the computer and software industries themselves and predates the Internet era of the 1990s [20]. There is no standard taxonomy of electronic commerce technologies, but they do include major categories such as magnetic ink character recognition, automatic teller machines, electronic funds transfer, stock market automation, facsimiles, email, point of sale systems, Internet service providers and electronic data interchange, as well as electronic retail trade and shopping websites [20].

3.2 Second-Generation Electronic Commerce

Second-generation or transistorized computers were associated with electronic commerce technologies such as magnetic ink character recognition or MICR

D.F. RICO *ET AL.*

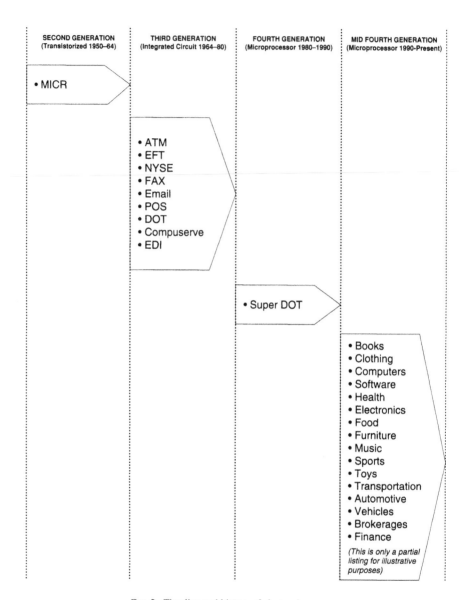

| SECOND GENERATION (Transistorized 1950–64) | THIRD GENERATION (Integrated Circuit 1964–80) | FOURTH GENERATION (Microprocessor 1980–1990) | MID FOURTH GENERATION (Microprocessor 1990-Present) |

• MICR

• ATM
• EFT
• NYSE
• FAX
• Email
• POS
• DOT
• Compuserve
• EDI

• Super DOT

• Books
• Clothing
• Computers
• Software
• Health
• Electronics
• Food
• Furniture
• Music
• Sports
• Toys
• Transportation
• Automotive
• Vehicles
• Brokerages
• Finance

(This is only a partial listing for illustrative purposes)

FIG. 2. Timeline and history of electronic commerce.

created in 1956, which was a method of 'encoding checks and enabling them to be sorted and processed automatically' [22].

3.3 Third-Generation Electronic Commerce

Third-generation or integrated-circuit-based computers were associated with electronic commerce technologies such as automatic teller machines, electronic funds transfer, stock market automation, facsimiles, email, point of sale systems, electronic bulletin boards and electronic data interchange. In 1965, automated teller machines were created [23], which were electronic machines or computers that automatically dispense money or cash [22]. In 1966, electronic funds transfer or EFT was created [24], which was 'a set of processes that substitutes electronic messages for checks and other tangible payment mechanisms' [22]. Also in 1966, the New York Stock Exchange or NYSE was first automated [25]. In 1971, facsimiles were created [26]. In 1973, email was created [19]. In 1975, electronic point of sale systems were created [27], which involved 'the collection in real-time at the point of sale, and storing in a computer file, of sales and other related data by means of a number of electronic devices' [28]. In 1976, the designated order turn-around or DOT was created, which automated small-volume individual trades [25]. In 1979, Compuserve launched one of the first electronic bulletin boards [19]. Also in 1979, electronic data interchange was created [29], which is the 'electronic movement of information, such as payments and invoices, between buyers and sellers' [30].

3.4 Fourth-Generation Electronic Commerce

Fourth-generation or microprocessor-based computers were associated with electronic commerce technologies such as the vast automation of the stock market. In 1984, the super designated order transfer 250 was launched to enable large-scale automatic program trading [25].

3.5 Mid-Fourth-Generation Electronic Commerce

Mid-fourth-generation computers were associated with electronic commerce technologies such as selected electronic services, electronic retail trade and electronic shopping and mail order houses. Selected electronic services consisted of industry sectors such as transportation and warehousing; information, finance, rental and leasing services; professional, scientific and technical services; administrative and support services; waste management and remediation services; health care and social assistance services; arts, entertainment and recreation services; accommodation and food services; and other services [21]. Electronic retail trade consisted of industry sectors such as motor vehicles and parts dealers; furniture and home furnishing stores; electronics and appliance stores; building materials, garden equipment and supplies stores; food and beverage stores; health and personal services; gasoline stations; clothing and accessories stores; sporting goods, hobby, book and music stores; general merchandise stores; miscellaneous store retailers; and non-store retailers [21]. And, electronic shopping and mail order houses consisted of industry sectors such as books and magazines; clothing and clothing accessories; computer hardware; computer software; drugs, health aids and beauty aids; electronics and appliances; food, beer and wine; furniture; music and videos; office equipment; sporting goods, toys, hobby goods, and games; other merchandise; and non-merchandise receipts [21].

Today, the U.S. electronic commerce industry garners revenues in excess of $2.4 trillion per year [31]. About $2.2 trillion is acquired from business-to-business or B2B commerce, also known as electronic data interchange or EDI. A good example of B2B is Wal-Mart computers, which automatically initiates an order and shipment from a wholesaler such as Proctor and Gamble when supplies run low. About $136 billion to $189 billion worth of U.S. electronic commerce comes from business-to-consumer (B2C), also known as online retail sales or Internet retailers. The best example of B2C is a consumer shopping for and ordering a textbook from Amazon. In 2007, 147 million Internet shoppers conducted 632.5 million transactions worth $136 billion to $189 billion. In total, there are 1.25 billion Internet users, the number of websites has reached 136 million and the number of web hosts has reached 470 million. Information technology, primarily in the form of the Internet contributes to more than 50% of total labour productivity growth in the top 10 industrialized nations and nearly 100% in China and India.

In 2006, the top 100 Internet retailers grew at an average rate of 19%, the bottom 100 from the top 500 grew at an average rate of 23%, startups grew at 55%, the fastest Internet retailers grew at 200%, and one firm grew at a rate of 400%. As many as 45% of U.S. Internet retailers are considered 'pure-plays'; that is, non-brick-and-mortar retailers such as Amazon. Traditional retailers such as Wal-Mart, Sears and others lag behind pure-plays in growth, conversion rates, customer satisfaction, website satisfaction and website quality. (Conversion rates refer to the percentage of Internet shoppers who make a purchase after visiting an electronic commerce website.) It's important to note that Internet retailing only garners about 3% to 4% of all retail sales in the U.S. That is, for every dollar spent by the Americans, only four cents is spent making online purchases. As the number of world-wide Internet users, shoppers and sales increase, this will result in unprecedented demands on the number of websites that need to be produced.

4. History of Software Methods

4.1 Database Design

One of the earliest software methods that emerged in the mainframe era of the 1960s was database design. As shown in Figures 3 and 4, database design is a process of developing the structure and organization of an information repository [32]. And, the U.S. formed a standard information resource dictionary system or IRDS [33], which is a 'logically centralized repository of data about all relevant information resources within an organization, often referred to as metadata' [34]. The use of flat files for the design of information repositories was one of the earliest forms of database design [35]. Network databases were one of the earliest forms of industrial-strength information repositories consisting of many-to-many relationships between entities or data records, examples of which include IBM's Information Management System or IMS/360 and UNIVAC's Data Management System or DMS 1100 [36]. Hierarchical databases soon emerged with a focus on organizing data into tree-like structures, which were believed to mimic the natural order of data in the real world [37]. Relational database design was introduced to create more reliable and less redundant information repositories based on the mathematical theory of sets [38].

4.2 Automatic Programming

Another of the earliest software methods that emerged in the mainframe era of the 1960s was automatic programming, which is also known as fourth-generation programming languages or 4GLs. Automatic programming is defined as the 'process and technology for obtaining an operational program and associated data structures automatically or semi-automatically, starting with only a high-level user-oriented specification of the environment and the tasks to be performed by the computer' [39]. Decision tables were one of the first automatic programming methods, which provided a simple format enabling both users and analysts to design computer software without any programming knowledge [40]. Programming questionnaires were also one of the first automatic programming methods, which provided an English-like yes or no questionnaire enabling non-computer programmers to answer a few discrete questions about their needs, leading to the automatic production of computer software [41]. The next evolution in automatic programming methods that emerged in the early midrange era was problem statement languages, characterized by the information system design and optimization system or ISDOS, which provided a means for users and other non-programmers to specify their needs and requirements and 'what' to do, without specifying 'how' the computer programs would perform their functions [42]. Special purpose languages also began emerging in the early midrange area, which

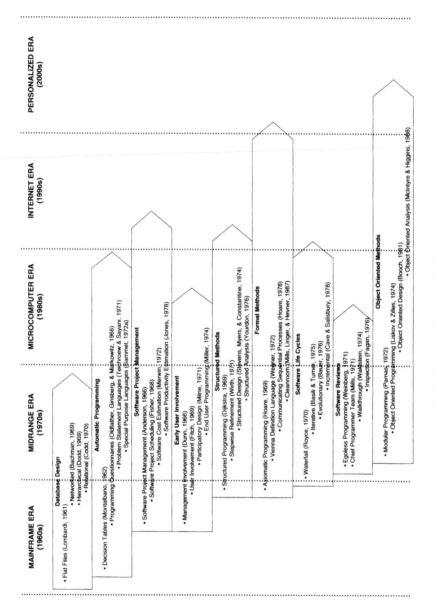

FIG. 3. Timeline and history of software methods.

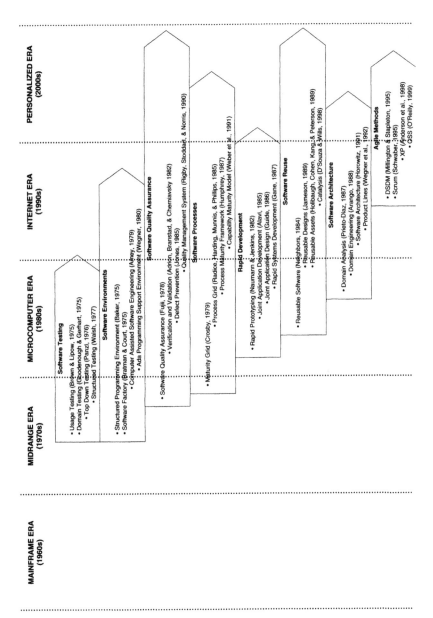

FIG. 4. Timeline and history of software methods (continued).

were regarded as very high level, English-like computer programming languages used for rapid prototyping and quick software composition, major examples of which include statistical analysis packages, mathematical programming packages, simplified database query languages and report generators [43].

4.3   Software Project Management

The earliest notions of software project management also emerged in the mainframe era of the 1960s. An early definition of software project management was the 'skillful integration of software technology, economics, and human relations' [44]. The project evaluation and scheduling technique or PEST was one of the first complete approaches to software project management emerging in this era [45]. Project network diagrams in the form of the program evaluation review technique or PERT and the critical path method or CPM, though not originating in computer programming, were soon applied for planning, scheduling and managing resources associated with software projects [46]. Cost-estimation techniques were soon added to the repertoire of software project management, especially for managing large U.S. military projects [47]. The framework for software project management was finally in place for years to come when basic measures of software productivity and quality were devised [48].

4.4   Early User Involvement

Early user involvement has long been recognized as a critical success factor in software projects since the earliest days of the mainframe era. Early user involvement is defined as 'participation in the system development process by representatives of the target user group' [49]. While project overruns were considered a normal part of the early computer age, scholars began calling for management participation to stem project overruns, which are now regarded as a 'management information crisis' [50]. By the late 1960s, 'user involvement' began to supplant management participation as a key for successfully designing software systems [51]. In the early 1970s, end users were asked to help design software systems themselves in what was known as 'participatory design' [52]. End user development quickly evolved from these concepts, which asked the end users to develop the applications themselves to help address the productivity paradox [53].

4.5   Structured Methods

The late mainframe period gave rise to structured methods as some of the earliest principles of software engineering to help overcome the software crisis. Structured methods are approaches for functionally decomposing software designs, e.g., expressing software designs in high-level components, which are further refined in terms of lower level components [54]. Structured programming emerged in this time-frame to help programmers create well-structured computer programs [55]. The next innovation in structured methods was called 'top down stepwise refinement', which consisted of the hierarchical design and decomposition of computer programs [56]. Structured design quickly followed suit, which is defined as 'a set of proposed general program design considerations and techniques for making coding, debugging, and modification easier, faster, and less expensive by reducing complexity' [57]. Structured analysis methods rounded out this family of methods by suggesting the use of graphs for depicting the decomposition of software functions and requirements [58].

4.6   Formal Methods

The late mainframe period also gave rise to formal methods, which would be used as the theoretical basis for software engineering for the next two decades. Formal methods are 'mathematically based languages, techniques, and tools for specifying and verifying' reliable and complex software systems [59]. Axiomatic programming is one of the first recognized formal methods, which uses mathematical set theories to design functionally correct software [60]. An entire set of formal semantics was soon devised to serve as a basis for creating mathematically correct computer programming languages called the Vienna definition language [61]. The communicating sequential processes method was then created by Tony Hoare to help design mathematically correct multi-tasking software systems [62]. The cleanroom or box-structured methodology was created to serve as a stepwise refinement and verification process for creating software designs [63]. Formal methods, primarily due the difficulty associated with their mathematical rigor, never enjoyed widespread adoption by the growing community of computer programmers [64].

4.7   Software Life Cycles

One of the first methods to come out of the early midrange era was the notion of software life cycles. A software life cycle is a 'collection of tools, techniques, and methods, which provide roles and guidelines for ordering and controlling the actions and decisions of project participants' [65]. The waterfall is one of the first recognized software life cycles consisting of seven stages: system requirements, software requirements, analysis, program design, coding, testing and operations [66] as popularized by Barry Boehm. The iterative software lifecycle appeared around the middle of the decade, which consisted of using a planned sequence of programming enhancements until computer software was complete [67]. The evolutionary software life cycle soon formed with notions of gradually enhancing computer programs rather than developing them in phases or iterations [68]. The incremental software life cycle followed next, recommending project assessments at each major milestone in order to identify and reduce risk [69]. The spiral model called for risk analysis between major milestones and prototypes as well [70].

4.8   Software Reviews

Software reviews emerged as a methodology in the very earliest stages of the midrange era. Software reviews are meetings held by groups of software engineers to review software products to identify and remove their defects [73]. Egoless programming was introduced in the early midrange era as a method of transforming software development from an individual craft into a loosely structured group activity [74]. Chief programmer teams emerged shortly thereafter to formalize the notion of egoless programming with one small difference; the team would have a clear leader [75]. Structured walkthroughs were quickly formed to once again place the responsibility for maintaining the overall program quality in the hands of the team, rather than a in the hands of a single individual [76]. Software inspections crystallized the concept of structured walkthroughs with a rigid meeting protocol for group reviews in order to optimize team performance [77]. In the same year, the U.S. military formed a standard with system design reviews, software specification reviews, preliminary design reviews, critical design reviews, test readiness reviews, functional configuration audits, physical configuration audits, formal qualification reviews and production readiness reviews [78].

4.9   Object-Oriented Methods

Object-oriented methods emerged in the midrange era as a direct response to calls for a software engineering discipline to mitigate the software crisis. Object-oriented methods are processes that 'allow a designer to generate an initial design model in the context of the problem itself, rather than requiring a mapping onto traditional computer science constructs' [79]. The Simula programming language actually emerged in the late 1960s, which was recognized by some as having modular and object-oriented programming features and capabilities [80]. Smalltalk soon followed as an offshoot of the flex programming language and reactive engine, which also had many early modular and object-oriented programming features [81]. However, the principles of modular and object-oriented programming – information hiding, self-contained data structures, co-located subroutines, and well-defined interfaces – were first formalized just after the emergence of Simula and Smalltalk programming languages, the principles of which were claimed to improve efficiency, flexibility and maintainability [84].

Object-oriented design emerged in the early microcomputer era to demonstrate one of the first graphical notations for describing object-oriented programming languages [85]. Finally, object-oriented analysis methods emerged, often reusing the tools of structured analysis to begin constructing specifications of software systems prior to devising their object oriented design [86].

4.10   Software Testing

Software testing gained recognition in the middle of the midrange era though system and hardware component testing had been the norm for at least two decades. Software testing is defined as 'the process of executing a software system to determine whether it matches its specification and executes in its intended environment' [87]. Software usage testing was developed based on the notion that software reliability could be improved by specifying how computer programs will be used, devising tests to model how users operate programs and then measuring the outcome of the testing [88]. Domain testing emerged at the same time with its principles of identifying test cases from program requirements, specifying a complete set of inputs using mathematical set theory and using set theory itself to prove program correctness when necessary [89]. Soon thereafter, top down testing was introduced, which recommended a unique test procedure for each software subroutine [90]. Finally, structured testing emerged as an approach to encapsulate best practices in software testing for novice computer programmers [91].

4.11   Software Environments

Software environments emerged in the middle of the midrange era as a means of improving software quality and productivity through automation. A software environment may be described as an 'operating system environment and a collection of tools or subroutines' [92]. A slightly better definition of software environment is a 'coordinated collection of software tools organized to support some approach to software development or conform to some software process model' [93], where software tools are defined as 'computer programs that assist engineers with the design and development of computer-based systems' [94]. Structured programming environments were created as a means of improving software reliability and productivity using guidelines, code libraries, structured coding, top down development, chief programmer teams, standards, procedures, documentation, education and metrics [95]. Software factories were soon created to introduce discipline and repeatability, software visualization tools, the capture of customer needs or requirements, automated software testing and software reuse [96]. Computer-assisted software engineering or CASE was also created to enhance software productivity and reliability by automating document production, diagram design, code compilation, software testing, configuration management, management reporting and sharing of data by multiple developers [97]. The Ada programming support environment or APSE was suggested as a core set of programming tools consisting of editors, compilers, debuggers, linkers, command languages and configuration management utilities [98]. Computer-aided software engineering was created to automate the tasks of documenting customer requirements, creating software architectures and designs, maintaining requirements traceability and configuration management [99]. Integrated computer-aided software engineering or I-CASE tools emerged, merging analysis and code generation tools [100]. However, I-CASE was a concept readily adopted by the U.S. DoD that was soon abandoned since it was never delivered and never worked well [101].

4.12   Software Quality Assurance

The modern day tenets of software quality assurance began to assume their current form in the late midrange era. Software quality assurance is defined as a 'planned and systematic pattern of all actions necessary to provide adequate confidence that the software conforms to established technical requirements' [102]. Software quality assurance was created to establish 'adherence to coding standards and conventions, compliance with documentation requirements and standards, and successful completion of activities' [103]. Software verification and validation was created to determine the adequacy of software requirements, software designs, software source code and regression testing during software maintenance [104]. Defect prevention was a structured process of determining the root causes of software defects and then institutionalizing measures to prevent their recurrence [105]. Quality management systems consisted of a set of organizational policies and procedures to ensure that the software satisfied its requirements [106].

4.13   Software Processes

Software processes were formed in the microcomputer era, though they were rooted in the traditions of software engineering, structured methods, software life cycles and software environments dating back to the late mainframe and early midrange eras. A software process is the 'collection of related activities seen as a coherent process subject to reasoning involved in the production of a software system' [107]. The maturity grid (e.g., uncertainty, awakening, enlightenment, wisdom and certainty), though not for software, inspired the software process modelling movement of the microcomputer era [108]. IBM then created its own process grid (e.g., traditional, awareness, knowledge, skill and wisdom and integrated management system) for conducting site studies of computer programming laboratories [109]. The process maturity framework was directly adapted from IBM's process grid [110], which was finally turned into the capability maturity model for U.S. military use [111].

4.14 Rapid Development

Rapid development was formalized in the microcomputer era though its tenets can be traced back to early notions of structured methods and prototyping. Rapid development is defined as a 'methodology and class of tools for speedy object development, graphical user interfaces, and reusable code for client-server applications' [112]. Rapid development leveraged productivity-enhancing technologies to build early models of information systems and involved end users in the definition of system requirements and designs [112]. Rapid development sought to control cost and schedule overruns; improve user acceptance, customer satisfaction, system quality and system success; and ultimately reduce information systems backlogs [112]. Rapid prototyping was defined as a process of quickly creating an informal model of a software system, soliciting user feedback and then evolving the model until it satisfied the complete set of customer requirements [113]. Joint application development was a process of having professional software developers assist end users with the development of their applications by evaluating their prototypes [114]. Joint application design, on the other hand, was a structured meeting between end users and software developers, with the objective of developing software designs that satisfied their needs [115]. Rapid systems development was an extension of the joint application design method, which advocated specific technological solutions such as relational databases and the completion of the software system, not just its design [116]. In a close adaptation, rapid application development recommended iterative rapid system development cycles in 60- to 120-day intervals [117].

4.15 Software Reuse

Software reuse assumed its current form in the early microcomputer era, though its earliest tenets can clearly be seen in literature throughout the 1950s, 1960s and 1970s. Software reuse is the 'use of existing software or software knowledge to construct new software' [118]. Software reuse was proposed as early as 1968 in order to help alleviate the 'software crisis' characterized by an explosion in computers and software complexity through the production of mass-produced software components [119]. The purpose of software reuse has evolved over the years to include improvements in productivity [120], reliability [121], quality [122] and cost efficiency [123]. Reusable software became synonymous with the Ada programming language in the

1980s though it was prophesied as a major strategy in 1968 and was a central management facet of Japanese software factories in the 1970s [124]. Toward the end of the microcomputer era, reusable software designs were considered just as important as reusable software source code [125]. In the same year, reusability was expanded to include requirements, designs, code, tests, and documents and dubbed 'reusable assets' [126]. Toward the end of the Internet era, catalysis was formed based on composing new applications from existing ones [127]. Though several studies chronicled software reuse success stories from the 1970s, 1980s and 1990s, [9, 128], scholars have concluded that software reuse has only had marginal success since 1968 [131].

4.16   Software Architecture

Software architecture began to assume a strategic role for managing the development of software systems near the end of the microcomputer era. Software architecture is defined as 'the structure and organization by which modern system components and subsystems interact to form systems and the properties of systems that can best be designed and analysed at the system level' [133]. Domain analysis was the discipline of identifying, capturing and organizing all of the information necessary to create a new software system [134]. Domain engineering was a process of managing reusable information about specific types of software systems, gathering architectural data and gathering data about the computer programs themselves [135]. Software architecture was a discipline of creating flexible software designs that were adaptable to multiple computer systems in order to respond to the rapidly changing military threats [136]. Software product lines soon emerged with an emphasis on evolving software architectures to reduce costs and risks associated with changes in design [137], along with software product families [138].

4.17   Agile Methods

Agile methods gained prominence in the late Internet and early personalized eras in part to accommodate the uniquely flexible nature of Internet technologies [139]. More to the point, the use of agile methods for Internet software was a reaction to the rise of traditional software development methods, which were too cumbersome, expensive, rigid and fraught with failure [139]. Downsizing was the norm and traditional methods were being used by large corporations in decline [139]. Agile methods are an approach for managing the development of software, which are based upon obtaining early customer feedback on a large number of frequent software releases [140]. In 2001, the 'agile manifesto' was created to outline the values and principles of agile methods and how they differed from traditional ones [141]. A council of 17 experts in agile methods met in order to find an 'alternative to documentation-driven, heavyweight software development processes'. They believed that 'in order to succeed in the new economy, to move aggressively into the era of e-business, e-commerce, and the web, companies have to rid themselves of their Dilbert manifestations of make-work and arcane policies'. Once the ground rules and assumptions of agile methods were established, they were able to get on with the business of writing the agile manifesto itself and publish it on the Internet. The agile manifesto began with the following statement: 'we are uncovering better ways of developing software by doing it and helping others do it'. Then the agile manifesto laid out four broad values: (a) 'working software over comprehensive documentation', (b) 'customer collaboration over contract negotiation', (c) 'individuals and interactions over processes and tools', and (d) 'responding to change over following a plan'. The agile manifesto itself was derived [139] from the dynamic system development methodology [142], scrum [143], extreme programming [144], open-source software development [145], crystal methods [146], feature-driven development [147], rational unified process [148], adaptive software development [149] and lean development [150]. Other approaches also influenced the development of agile methods such as the new product development game, new-product development rhythm, synch-n-stabilize, judo strategy and Internet time, which will be described later [139].

The dynamic system-development methodology or DSDM has three broad phases, which consist of requirement prototypes, design prototypes and an implementation or production phase [142]. Scrum is a light-weight software-development process consisting of implementing a small number of customer requirements in two- to four-week sprint cycles [143]. Extreme programming or XP consists of collecting informal requirements from on-site customers, organizing teams of pair programmers, developing simple designs, conducting rigorous unit testing, and delivering small and simple software packages in short two-week intervals [144]. Open-source software development involves freely sharing, peer reviewing, and rapidly evolving software source code for the purpose of increasing its quality and reliability [145]. Crystal methods involve frequent delivery; reflective improvement; close communication; personal safety; focus; easy access to expert users; and a technical environment with automated testing, configuration management and frequent integration [146]. Feature-driven development involves developing an overall model, building a features list, planning by feature, designing by feature and building by feature [147]. The rational unified process involves a project management, business modelling, requirements, analysis and design, implementation, test, configuration management, environment and deployment workflow [148]. Adaptive software development involves product initiation, adaptive cycle planning, concurrent feature development, quality review and final quality assurance and release [149]. And, lean development involves eliminating waste, amplifying the learning process, making decisions as late as possible, delivering products as fast as possible, empowering teams, building integrity in, and seeing systems as a whole [150]. Agile methods such as extreme programming adapted customer feedback, iterative development, well-structured teams, flexibility from the new product development game, new product development rhythm, synch-n-stabilize, judo strategy and Internet time [139].

5. History of Software Quality Measurement

5.1   Software Size

One of the earliest known measures used to describe computer programs was software size [151]. Software size is a measure of the volume, length, quantity, amount and overall magnitude of a computer program [152]. In the mid 1960s, lines of code or LOC was one of the first known measures of software size, which referred to the number of computer instructions or source statements comprising a computer program and is usually expressed as thousands of lines of code [153]. Almost a decade later in the mid to late 1970s, more sophisticated measures of software size emerged such as token count, volume, function count and function points [152]. Recognizing that individual lines of code had variable lengths, token count was created to distinguish between unequally sized lines of code, which was technically defined as 'basic syntactic units distinguishable by a compiler' [154]. In yet another attempt to accurately gauge the size of an individual line of code, volume was created to measure the actual size of a line of code in bits, otherwise known as binary zeros and ones [154]. Shortly thereafter, function count was created to measure software size in terms of the number of modules or subroutines [155]. Function points was another major measure of software size, which was based on estimating the number of inputs, outputs, master files, inquiries and interfaces [156]. Though software size is not a measure of software quality itself, it formed the basis of many measures or ratios of software quality right on through the modern era (e.g., number of defects, faults, or failures per line of code or function point). Furthermore, some treatises on software metrics consider software size to be one of the most basic measures of software complexity [157]. Thus, a history of software quality measurement may not be complete without the introduction of an elementary discussion of software size.

5.2   Software Errors

One of the earliest approaches for measuring software quality was the practice of counting software errors dating back to the 1950s when digital computers emerged. Software errors are human actions resulting in defects, defects sometimes manifest

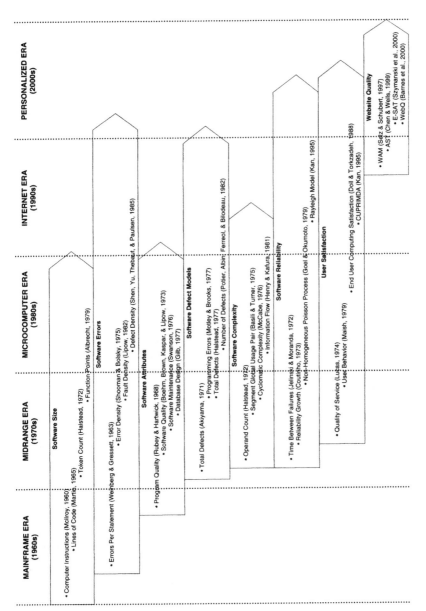

Fig. 5. Timeline and history of software quality measures.

themselves as faults, and faults lead to failures, which are often referred to as software crashes [157]. The concept of 'errors per statement' first appeared in the early 1960s [158] and studies of 'error proneness' intensified towards the end of the decade [159]. The term error density was coined in the mid 1970s, which referred to the simple ratio of errors to software size [160]. Fault density was also a measure of software quality, which referred to the ratio of anomaly-causing faults to software size [161]. The term defect density subsumed the measure of error and fault density in the mid 1980s, which referred to the ratio of software errors to software size [162]. Many unique types of errors were counted, such as number of requirement, design, coding, testing and maintenance errors, along with number of changes and number of changed lines of code [152]. Even the term problem density emerged in the early 1990s, which referred to the number of problems encountered by customers to measure and track software quality [157]. The practice of counting errors, defects, faults and failures as a means of measuring software quality enjoyed widespread popularity for more than five decades.

5.3   Software Attributes

Another of the earliest approaches for measuring software quality was the practice of quantifying and assessing attributes or characteristics of computer programs. Software attributes are an 'inherent, possibly accidental trait, quality or property' such as functionality, performance or usability [163]. Logicon designed a model to measure software attributes such as correctness, logicality, non-interference, optimizability, intelligibility, modifiability and usability [164]. Next, TRW identified software attributes such as portability, reliability, efficiency, modifiability, testability, human engineering and understandability [165]. These early works led to numerous specialized spin-offs such as a framework for measuring the attributes of software maintenance [166] and even a database's design [167]. Spin-offs continued to emerge with an increasing focus on operationalizing these attributes with real software metrics [168]. By the mid 1980s, this practice reached Japan [171] and a comprehensive framework emerged replete with detailed software measures [172]. Use of software attributes to measure software quality was exemplified by the functionality, usability, reliability, performance, and supportability, which was named the FURPS model [173]. Software attributes enjoyed widespread use among practitioners throughout the 1970s and 1980s because of their simplicity, though scientists favoured statistical models.

5.4   Static Defect Models

One of the earliest approaches for predicting software quality was the use of statistical models referred to as static reliability or static software defect models.

'A static model uses other attributes of the project or program modules to estimate the number of defects in software' [157], ignoring 'rate of change' [152]. One of the earliest software defect models predicted the number of defects in a computer program as a function of size, decision count or number of subroutine calls [174]. Multi-linear models were created with up to 10 inputs for the various types of statements found in software code such as comments, data and executable instructions [175]. The theory of software science was extended to include defect models by using volume as an input, which itself was a function of program language and statement length [154]. Research on software defect models continued with more extensions based on software science, cyclomatic complexity, path and reachability metrics [176]. More defect models were created by mixing defects, problems and software science measures such as vocabulary, length, volume, difficulty and effort [162]. Later, IBM developed models for predicting problems, fielded defects, arrival rate of problems and backlog projection, which were used to design midrange operating systems [157]. Static linear or multi-linear statistical models to predict defects continue to be useful tools well into modern times, though older dynamic statistical reliability models are overtaking them.

5.5 Software Complexity

With its emergence in the early 1970s, the study of software complexity became one of the most common approaches for measuring the quality of computer programs. Software complexity is defined as 'looking into the internal dynamics of the design and code of software from the program or module level to provide clues about program quality' [157]. Software complexity sprang from fervor among research scientists eager to transform computer programming from an art into a mathematically based engineering discipline [177]. Many technological breakthroughs in the two decades prior to mid 1970s led to the formation of software complexity measures. These included the advent of digital computers in the 1950s, discovery of high-level computer programming languages and the formation of compiler theory. Furthermore, flowcharting was routinely automated, axiomatic theorems were used for designing new computer languages and analysis of numerical computer algorithms became commonplace. As a result, three major classes of software complexity metrics arose for measuring the quality of software: (a) data structure, (b) logic structure and (c) composite metrics [178]. One of the first data structure metrics was the count of operands, which measured the number of variables, constants and labels in a computer program versus measuring logic [177]. The segment-global-usage-pair metric determined complexity by counting references to global variables, a high number of which was considered bad among coders [67]. Another data structure metric was the span between variables, which measured how many logic structure statements existed between variables where a higher number was poor [179]. A unique data structure metric for measuring software quality was the number of live variables within a procedure or subroutine as a sign of undue complexity [180]. One data structure metric surviving to modern times is the information flow, or fan in-fan out metric, which measures the number of modules that exchange data [181]. Logic structure metrics were cyclomatic complexity or paths [182], minimum paths [183] and gotos or knots [184]. Also included were nesting [185], reachability [186], nest depth [187] and decisions [162]. Composite metrics combined cyclomatic complexity with other attributes of computer programs to achieve an accurate estimate of software quality [188]. They also included system complexity [191] and syntactic construct [192]. Finally, it's important to note that most complexity metrics are now defunct, though cyclomatic complexity, which arose out of this era, is still used as a measure of software quality today.

5.6   Software Reliability

Software reliability emerged in the early 1970s and was created to predict the number of defects or faults in software as a method of measuring software quality. Software reliability is the 'probability that the software will execute for a particular period of time without failure, weighted by the cost to the user of each failure encountered' [193]. Major types of reliability models include: (a) finite versus infinite failure models [194], (b) static versus dynamic [152] and (c) deterministic versus probabilistic [195]. Major types of dynamic reliability models include: life cycle versus reliability growth [157] and failure rate, curve fitting, reliability growth, nonhomogeneous Poisson process and Markov structure [195]. One of the first and most basic failure rate models estimated the mean time between failures [196]. A slightly more sophisticated failure rate model was created based on the notion that software became more reliable with the repair of each successive code failure [197]. The next failure rate model assumed that the failure rate was initially constant and then began to decrease [198]. Multiple failure rate models appeared throughout the 1970s to round out this family of reliability models [199]. Reliability or 'exponential' growth models followed the emergence of failure rate models, which measured the reliability of computer programs during testing as a function of time or the number of tests [202, 203]. Another major family of reliability models is the non-homogeneous Poisson process models, which estimate the mean number of cumulative failures up to a certain point in time [205]. Reliability models estimate the number of software failures after development based on failures encountered during testing and operation. Though rarely mentioned, the Rayleigh life cycle reliability model accurately estimates defects inserted and removed throughout the software lifecycle [157]. Some researchers believed that the use of software reliability models offered the best hope for transforming computer programming from a craft industry into a true engineering discipline.

5.7   User Satisfaction

User satisfaction gradually became a measure of software quality during the 1950s, 1960s and 1970s [208] User satisfaction is defined as 'the sum of one's feelings or attitudes toward a variety of factors affecting that situation', e.g., computer use and adoption by end users [212]. Though not the first, one study of user satisfaction analysed attitudes towards quality of service, management support, user participation, communication and computer potential [213]. A more complex study of user satisfaction considered the feelings about staff, management support, preparation for its use, access to system, usefulness, ease of use and flexibility [214]. Most studies until 1980 focused on the end user's satisfaction with regards to software developers; but one study squarely focused on the end user's satisfaction with regards to the software itself [215]. One of the first studies to address a variety of software attributes such as software accuracy, timeliness, precision, reliability, currency and flexibility appeared [216]. Studies throughout the 1980s addressed user satisfaction with both designers and software [212, 217]. The late 1980s marked a turning point, with studies focusing entirely on user satisfaction with the software itself and attributes such as content, accuracy, format, ease of use and timeliness of the software [221]. A study of user satisfaction at IBM was based on reliability, capability, usability, installability, maintainability, performance and documentation factors [222]. Throughout the 1990s, IBM used a family of user satisfaction models called UPRIMD, UPRIMDA, CUPRIMDA and CUPRIMDSO, which referred differently to factors of capability, usability, performance, reliability, installability, maintainability, documentation, availability, service and overall satisfaction [157]. User satisfaction, now commonly referred to as customer satisfaction, is undoubtedly related to earlier measures of software attributes, usability or user friendliness of software and more recently web quality.

5.8   Website Quality

With their emergence in the late 1990s, following the user satisfaction movement, models of website quality appeared as important measures of software quality [223]. One of the first models of website quality identified background, image size, sound file display and celebrity endorsement as important factors of software quality [224]. The web assessment method or WAM quickly followed with quality factors of external bundling, generic services, customer-specific services and emotional experience [225]. In what promised to be the most prominent web quality model, attitude towards the site or AST had quality factors of entertainment, informativeness, and organization [226]. The next major model was the e-satisfaction model with its five factors of convenience, product offerings, product information, website design and financial security [227]. The website quality model or WebQual for business school portals was based on factors of ease of use, experience, information and communication and integration [228]. An adaptation of the service quality or ServQual model, WebQual 2.0 measured quality factors such as tangibles, reliability, responsiveness, assurance and empathy [229]. The electronic commerce user consumer satisfaction index or ECUSI consisted of 10 factors such as product information, consumer service, purchase result and delivery, site design, purchasing process, product merchandising, delivery time and charge, payment methods, ease of use and additional information services [230]. On the basis of nine factors, the website quality or SiteQual model consisted of aesthetic design, competitive value, ease of use, clarity of ordering, corporate and brand equity, security, processing speed, product uniqueness and product quality assurance [231]. In what promised to be exclusively for websites, the Internet retail service quality or IRSQ model was based on nine factors of performance, access, security, sensation, information, satisfaction, word of mouth, likelihood of future purchases and likelihood of complaining [232]. In a rather complex approach, the expectation-disconfirmation effects on web-customer satisfaction or EDEWS model consists of three broad factors (e.g., information quality, system quality and web satisfaction) and nine sub-factors [233]. In one of the smallest and most reliable website quality models to date, the electronic commerce retail quality or EtailQ model consists of only four major factors (e.g., fulfillment and reliability, website design, privacy and security, and customer service) and only 14 instrument items [234]. On the basis of techniques for measuring software quality dating back to the late 1960s, more data have been collected and validated using models of website quality than any other measure.

6. History of Agile Methods

6.1 New Product Development Game

As shown in Figure 6, two management scholars from the School of International Corporate Strategy at Hitotsubashi University in Tokyo, Japan, published a management approach called the 'new product development game' in the Harvard Business Review in early 1986 [235]. In their article, they argued that Japanese 'companies are increasingly realizing that the old sequential approach to developing new products simply will not get the job done'. They cited the sport of Rugby as the inspiration for the principles of their new product development game – In particular, Rugby's

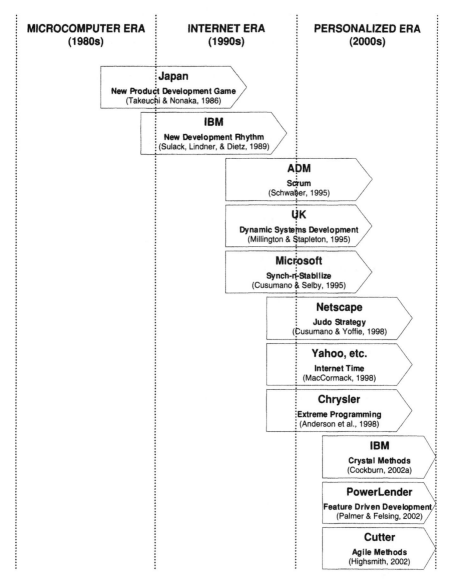

FIG. 6. Timeline and history of agile methods.

special play called the Scrum, when the players interlock themselves together as a tightly bound group to gain possession of the ball. The new product development game consisted of six major factors: (a) built-in instability, (b) self-organizing project teams, (c) overlapping development phases, (d) multi-learning, (e) subtle control and (f) organizational transfer of learning. They went on to demonstrate how four Japanese firms, e.g., Fuji-Xerox, Canon, Honda and NEC, applied the six factors of the new product development game to develop six major products, which became market successes. The six major factors of the new product development game were not unlike the total quality management and concurrent engineering movements that were popular in the U.S. during that timeframe, and their work inspired the development of agile methods for the next 20 years.

6.2   New Development Rhythm

In 1989, three managers from IBM in Rochester, Minnesota, published an article on how IBM devised a management approach called the 'new development rhythm', to bring the AS/400 midrange computer to market in only two years [236]. In their article, they stated that 'user involvement programs yielded a product offering that met the user requirements with a significantly reduced development cycle'. The new development rhythm consisted of six major factors: (a) modularized software designs, (b) software reuse, (c) rigorous software reviews and software testing, (d) iterative development, (e) overlapped software releases and (f) early user involvement and feedback. IBM's new development rhythm was a remarkable feat of management science and boasted a long list of accomplishments: (a) time-to-market improvement of 40%, (b) development of seven million lines of operating system software in 26 months, (c) compatibility with 30 billion lines of commercial applications, (d) $14 billion in revenues and (e) the IBM corporation's first Malcolm Baldrige National Quality Award. While there was nothing innovative about IBM's new development rhythm, it was IBM's audacity to apply these academic textbook approaches to commercial product development that was unique.

6.3   Scrum

In 1993, Jeff Sutherland of the Easel Corporation adapted the principles from the 'new product development game' [235] to the field of computer programming management, explicitly calling it 'scrum' [143]. In particular, scrum assumes that the 'systems development process is an unpredictable and complicated process that can only be roughly described as an overall progression'. Furthermore, scrum's creators believed 'the stated philosophy that systems development is a well-understood approach that can be planned, estimated, and successfully completed has proven incorrect in practice'. Therefore, scrum's creators set out to define a process as a 'loose set of activities that combines known, workable tools and techniques with the best that a development team can devise to build systems'. Today, scrum is composed of three broad phases: (a) pre-sprint planning, (b) sprint and (c) post-sprint meeting. During the pre-sprint planning phase, computer programmers gather to prioritize customer needs. During the sprint phase, computer programmers do whatever it takes to complete a working version of software that meets a small set of high-priority customer needs. Finally, during the post-sprint meeting, computer programmers demonstrate working software to their customers, adjust their priorities and repeat the cycle.

6.4   Dynamic Systems Development Method

In 1993, 16 academic and industry organizations in the United Kingdom banded together to create a management approach for commercial software called the 'dynamic systems development method' or simply DSDM [142]. Their goal was to 'develop and continuously evolve a public domain method for rapid application development' in an era dominated by proprietary methods. Initially, DSDM emphasized three success factors: (a) 'the end user community must have a committed senior staff that allows developers easy access to end users', (b) 'the development team must be stable and have well established skills' and (c) 'the application area must be commercial with flexible initial requirements and a clearly defined user group'. These success factors would later be expanded to include functionality versus quality, product versus process, rigorous configuration management, a focus on business objectives, rigorous software testing, risk management and flexible software requirements. DSDM consists of five major stages: (a) feasibility study, (b) business study, (c) functional model iteration, (d) design and build iteration and (e) implementation. The goal of DSDM is to explore customer requirements by building at least two full-scale prototypes before the final system is implemented.

6.5   Synch-N-Stabilize

In 1995, management scholars from MIT's Sloan School of Management and the University of California at Irvine published a textbook on how Microsoft managed the development of software for personal computers, dubbed as the 'sync-n-stabilize' approach [237]. The scholars were experts on software management approaches for the mainframe market and their two-year case study from 1993 to 1995 was a grounded theory or emergent research design, which led them to some startling conclusions. At one point in their textbook, they stated that 'during this initial research, it became clear why Microsoft was able to remain on top in its industry while most contemporaries from the founding years of the 1970s disappeared'. The synch-n-stabilize approach consisted of six major factors: (a) parallel programming and testing, (b) flexible software requirements, (c) daily operational builds, (d) iterative development, (e) early customer feedback and (f) use of small programming teams. Microsoft's success was indeed remarkable, and their synch-n-stabilize approach did indeed help them create more than 20 million lines of code for Windows and Office 95, achieve customer satisfaction levels of 95% and maintain annual profit margins of approximately 36%.

6.6   Judo Strategy

In 1998, two management scholars from both the Harvard Business School and MIT's Sloan School of Management published a textbook on how Netscape managed the development of software for the Internet, dubbed as the 'judo strategy' [238]. The scholars were experts on software management approaches for the personal computer market and their one year case study from 1997 to 1998 was a grounded theory or emergent research design, which prophetically led them to be critical of Netscape's future. Whereas Microsoft's strategic advantage was its immense intellectual capital, Netscape's only advantage seemed to be its first-mover status, which was quickly eroding to Microsoft's market share for browsers at the time their book was published. In fact, the authors criticized Netscape for not having a technical CEO in the fast-moving Internet market, which was a very unconventional view among management scholars. Some of the more notable factors characteristic of Netscape's judo strategy included: (a) design products with modularized architectures; (b) use parallel development; (c) rapidly adapt to changing market priorities; (d) apply as much rigorous testing as possible and (e) use beta testing and open source strategies to solicit early market feedback on features, capabilities, quality and architecture.

6.7   Internet Time

In 1998, a management scholar from the Harvard Business School conducted a study on how U.S. firms manage the development of websites, referring to his approach as 'Internet time' [239]. His study states that 'constructs that support a more flexible development process are associated with better performing projects'. Basically, what he did was survey 29 software projects from 15 Internet firms such as Microsoft, Netscape, Yahoo, Intuit and Altavista. He set out to test the theory that website quality was associated with three major factors: (a) greater investments in architectural design, (b) early market feedback and (c) greater amounts of generational experience. Harvard Business School scholars believed that firms must spend a significant amount of resources to create flexible software designs, they must incorporate customer feedback on working 'beta' versions of software into evolving software designs, and higher website quality will be associated with more experience among computer programmers. After determining the extent to which the 29 website software projects complied with these 'Internet time' factors through a process of interviews and surveys, he then assembled a panel of 14 industry experts to objectively evaluate the associated website quality. Statistical analysis supported two of the hypotheses, e.g., greater architectural resources and early market feedback were associated with higher website quality, but not the third, e.g., greater experience among computer programmers is associated with higher website quality. This was one of the first studies to offer evidence in support of agile methods.

6.8 Extreme Programming

In 1998, 20 software managers working for the Chrysler Corporation published an article on how they devised a management approached called 'extreme programming' or XP to turn around a failing software project that would provide payroll services for 86 000 Chrysler employees [144]. Today, extreme programming is synonymous with agile methods or agile programming and is one of the most widely used agile methods although its market lead is eroding. In their article, they stated that 'extreme programming rests on the values of simplicity, communication, testing, and aggressiveness'. They also stated that the 'project had been declared a failure and all code thrown away, but using the extreme programming methodology, Chrysler started over from scratch and delivered a very successful result'. Extreme programming consists of 13 factors: (a) planning game, (b) small releases, (c) metaphor, (d) simple design, (e) tests, (f) refactoring, (g) pair programming, (h) continuous integration, (i) collective ownership, (j) onsite customer, (k) 40 hour workweek, (l) open workspace and (m) just rules. The planning game consists of estimation of the scope and timing of releases. Small releases consist of groups of iterations that will be put into production when complete. Metaphors are a common nomenclature for objects and classes. Simple design is self-evident; that is keeping the software architecture and design as simple as possible without adding unnecessary bells and whistles. Tests are unit tests that must be written before the code is written and run after the code is complete. Refactoring is defined as the continuous refining of the designs and code for simplicity and efficiency. Pair programming means a team of two programmers responsible for writing a software module or group of modules. Continuous integration is also self-evident; all code must be integrated with the system soon after it is written and unit tested as a form of validation. Collective ownership means that anyone has the authority to redesign and recode any portion of the system. Onsite customer means that a customer is always present with the software team. Open workspace refers to collocated teams with few walls to optimize communication. And, just rules means programmers must agree to a common set of flexible rules. What these 20 software managers did was start over, get an informal statement of customer needs, gradually evolve a simple system design using iterative development, apply rigorous testing, use small teams of programmers, and get early customer feedback on their evolving design. In the end, Chrysler was able to deploy an operational payroll system serving more than 86 000 employees.

6.9   Crystal Methods

In 1991, a software manager with IBM was asked to create an approach for managing the development of object-oriented systems called 'crystal methods' [146]. Crystal methods were piloted on a '$15 million firm, fixed-price project consisting of 45 people'. Crystal methods are a 'family of methods with a common genetic code, one that emphasizes frequent delivery, close communication and reflective improvement'. Crystal methods are a family of 16 unique approaches for project teams ranging from 1 to 1000 people and project criticality ranging from loss of comfort to loss of life. The seven properties of crystal methods are: (a) frequent delivery; (b) reflective improvement; (c) close communication; (d) personal safety; (e) focus; (f) easy access to expert users and (g) a technical environment with automated testing, configuration management and frequent integration. The five strategies of crystal methods are: (a) exploratory 360, (b) early victory, (c) walking skeleton, (d) incremental re-architecture and (e) information radiators. The nine techniques of crystal methods are: (a) methodology shaping, (b) reflection workshop, (c) blitz planning, (d) Delphi estimation, (e) daily stand-ups, (f) agile interaction design, (g) process miniature, (h) side-by-side programming and (i) burn charts. The eight roles of crystal methods are: (a) sponsor, (b) team member, (c) coordinator, (d) business expert, (e) lead designer, (f) designer-programmer, (g) tester and (h) writer. The work products include a mission statement, team structure and conventions, reflection workshop results, project map, release plan, project status, risk list, iteration plan and status, viewing schedule, actor-goal list, use cases and requirements file, user role model, architecture description, screen drafts, common domain model, design sketches and notes, source code, migration code, tests, packaged system, bug reports and user help text.

6.10   Feature-Driven Development

In 1997, three software managers and five software developers created a software development approach called 'feature driven development' to help save a failed project for an international bank in Singapore [147]. In their textbook, they stated that 'the bank had already made one attempt at the project and failed, and the project had inherited a skeptical user community, wary upper management, and a demoralized

TABLE I
SUMMARY OF PRACTICES AND PROCESSES OF AGILE METHODS

Feature	FDD	Extreme programming	DSDM	Scrum
Practice	• Domain object modeling • Developing by feature • Class (code) ownership • Feature teams • Inspections • Regular build schedule • Configuration management • Reporting/visibility of results	• Planning game • Small releases • Metaphor • Simple design • Tests • Refactoring • Pair programming • Continuous integration • Collective ownership • On-site customer • 40-hour weeks • Open workspace • Just rules	• Active user involvement • Empowered teams • Frequent delivery • Fitness (simplicity) • Iterations and increments • Reversible changes • Baselined requirements • Integrated testing • Stakeholder collaboration	• Product backlog • Burndown chart • Sprint backlog • Iterations and increments • Self managed teams • Daily scrums
Process	**Develop an Overall Model** Form the Modeling Team Conduct a Domain Walkthrough Study Documents Develop Small Group Models Develop a Team Model Refine the Overall Object Model Write Model Notes Internal and External Assessment **Build a Features List** Form the Features List Team Build the Features List Internal and External Assessment	**User Stories** *Requirements* Acceptance Tests **Architectural Spike** System Metaphor **Release (1)** Release Planning *Release Plan* Iteration (1) Iteration Planning *Iteration Plan* Daily Standup Collective Code Ownership	**Feasibility Study** Feasibility Report Feasibility Prototype (optional) Outline Plan Risk Log **Business Study** Business Area Definition Prioritized Requirements List Development Plan System Architecture Definition Updated Risk Log **Functional Model Iteration** Functional Model	**Iteration (1)** Sprint Planning Meeting Product Backlog Sprint Backlog Sprint Daily Scrum Shippable Code Sprint Review Meeting Shippable Code Sprint Retrospective Meeting **Iteration (2)** Sprint Planning Meeting Product Backlog

continued

TABLE I
continued

Feature	FDD	Extreme programming	DSDM	Scrum
Process	**Plan by Feature** Form the Planning Team Determine Development Sequence Assign Features to Chief Coders Assign Classes to Developers Self Assessment **Iteration (1)** Design by Feature Form a Feature Team Conduct Domain Walkthrough Study Referenced Documents Develop Sequence Diagrams Refine the Object Model Write Class/Method Prologue Design Inspection Build by Feature Implement Classes/Methods Conduct Code Inspection Unit Test Promote to the Build **Iteration (2)** Design by Feature Form a Feature Team Conduct Domain Walkthrough Study Referenced Documents Develop Sequence Diagrams Refine the Object Model Write Class/Method Prologue Design Inspection	Create Unit Tests *Unit Tests* Pair Programming Move People Around Refactor Mercilessly Continuous Integration Acceptance Testing Iteration (2) Iteration Planning *Iteration Plan* Daily Standup Collective Code Ownership Create Unit Tests *Unit Tests* Pair Programming Move People Around Refactor Mercilessly Continuous Integration Acceptance Testing Iteration (n) Iteration Planning *Iteration Plan* Daily Standup Collective Code Ownership Create Unit Tests *Unit Tests* Pair Programming Move People Around Refactor Mercilessly	Functional Prototype (1) Functional Prototype Functional Prototype Records Functional Prototype (2) Functional Prototype Functional Prototype Records Functional Prototype (n) Functional Prototype Functional Prototype Records Non-functional Requirements List Functional Model Review Records Implementation Plan Timebox Plans Updated Risk Log **Design and Build Iteration** Timebox Plans Design Prototype (1) Design Prototype Design Prototype Records Design Prototype (2) Design Prototype Design Prototype Records Design Prototype (n) Design Prototype Design Prototype Records Tested System Test Records **Implementation** User Documentation	Sprint Backlog Sprint Daily Scrum Shippable Code Sprint Review Meeting Shippable Code Sprint Retrospective Meeting **Iteration (n)** Sprint Planning Meeting Product Backlog Sprint Backlog Sprint Daily Scrum Shippable Code Sprint Review Meeting Shippable Code Sprint Retrospective Meeting

Process

Build by Feature
Implement Classes/Methods
Conduct Code Inspection
Unit Test
Promote to the Build
Iteration (n)
Design by Feature
Form a Feature Team
Conduct Domain Walkthrough
Study Referenced Documents
Develop Sequence Diagrams
Refine the Object Model
Write Class/Method Prologue
Design Inspection
Build by Feature
Implement Classes/Methods
Conduct Code Inspection
Unit Test
Promote to the Build

Continuous Integration
Acceptance Testing
Release (2)
Iteration (1)
Iteration (2)
Iteration (n)
Release (n)
Iteration (1)
Iteration (2)
Iteration (n)

Trained User Population
Delivered System
Increment Review Document

development team'. Furthermore, they stated that 'the project was very ambitious, with a highly complex problem domain spanning three lines of business, from front office automation to backend legacy system integration'. In order to address this highly complex problem domain that had already experienced severe setbacks, they created an agile and adaptive software development process that is 'highly iterative, emphasizes quality at each step, delivers frequent tangible working results, provides accurate and meaningful progress, and is liked by clients, managers, and developers'. As shown in Table I, feature- driven development consists of five overall phases or processes: (a) develop an overall model, (b) build a features list, (c) plan by feature, (d) design by feature and (e) build by feature. Feature driven development also consists of other best practices in software management and development such as domain object modeling, developing by feature, individual class ownership, feature teams, inspections, regular builds, configuration management and reporting and visibility of results.

7. History of Studies on Agile Methods

7.1   Harvard Business School I

In 1998, two management scholars from the Harvard Business School conducted a survey of 391 respondents to test the effects of flexible versus inflexible product technologies, as shown in Figure 7 and Table II [240]. What they found was that projects using inflexible product technologies required over two times as much engineering effort as flexible product technologies (e.g., 17.94 vs. 8.15 months).

7.2   Harvard Business School II

In 1998, another management scholar from the Harvard Business School conducted a survey of 29 projects from 15 U.S. Internet firms to test the effects of flexible software development management approaches on website quality [239]. What he found was that flexible product architectures and customer feedback on early beta releases were correlated to higher levels of website quality.

7.3   Boston College Carroll School of Management

In 1999, two management scholars from Boston College's Carroll School of Management conducted a case study of 28 software projects to determine the effects of iterative development on project success [241]. What they found was that software projects that use iterative development deliver working software 38% sooner, complete their projects twice as fast, and satisfy over twice as many software requirements.

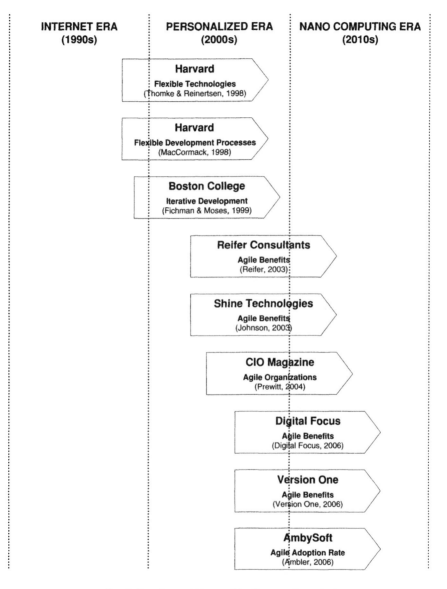

FIG. 7. Timeline and history of studies on agile methods.

7.4   Reifer Consultants

In 2003, Reifer Consultants conducted a survey of 78 projects from 18 firms to determine the effects of using agile methods to manage the development of software [242]. What they found was that 14% to 25% of respondents experienced productivity gains, 7% to 12% reported cost reductions and 25% to 80% reported time-to-market improvements.

7.5   Shine Technologies

In 2003, Shine Technologies conducted an international survey of 131 respondents to determine the effects of using agile methods to manage the development of software [243]. What they found was that 49% of the respondents experienced cost reductions, 93% of the respondents experienced productivity increases, 88% of the respondents experienced quality increases and 83% experienced customer satisfaction improvements.

7.6   CIO Magazine

In 2004, CIO Magazine conducted a survey of 100 information technology executives with an average annual budget of $270 million to determine the effects of agile management on organizational effectiveness [244]. What they found was that 28% of respondents had been using agile management methods since 2001, 85% of the respondents were undergoing enterprise-wide agile management initiatives, 43% of the respondents were using agile management to improve organizational growth and market share, and 85% said agile management was a core part of their organizational strategy.

7.7   Digital Focus

In 2006, Digital Focus conducted a survey of 136 respondents to determine the effects of using agile methods to manage the development of software [245]. What they found was that 27% of the respondents were adopting agile methods for a project, 23% of the respondents were adopting agile methods company wide, 51% of the respondents wanted to use agile methods to speed up the development process, 51% of the respondents said they lacked the skills necessary to implement agile methods at the project level, 62% of the respondents said they lacked the skills necessary to implement agile methods at the organization level and 60% planned on teaching themselves how to use agile methods.

7.8 Version One

In 2006, Version One conducted an international survey of 722 respondents to determine the effects of using agile methods to manage the development of software [246]. What they found was that 86% of the respondents reported time-to-market improvements, 87% of the respondents reported productivity improvements, 86% of the respondents reported quality improvements, 63% of the respondents reported cost reductions, 92% of the respondents reported the ability to manage changing priorities, 74% of the respondents reported improved morale, 72% of the respondents reported risk reductions, 66% of the respondents reported satisfaction of business goals and 40% were using the scrum method.

7.9 AmbySoft 2006

In 2006, Ambysoft conducted an international survey of 4232 respondents to determine the effects of using agile methods to manage the development of software [247]. What they found was that 41% of organizations were using agile methods; 65% used more than one type of agile method; 44% reported improvements in productivity, quality and cost reductions; and 38% reported improvements in customer satisfaction.

7.10 AmbySoft 2007

In 2007, Ambysoft conducted another international survey of 781 respondents to further determine the effects of using agile methods to manage the development of software [248]. What they found was that 69% of organizations had adopted agile methods, 89% of agile projects had a success rate of 50% or greater, and 99% of organizations are now using iterative development.

7.11 UMUC

In 2007, a student at the University of Maryland University College (UMUC) conducted a survey of 250 respondents to determine the effects of using agile methods on website quality [249]. What he found was that: (a) 70% of all developers are using many if not all aspects of agile methods; (b) 79% of all developers using agile methods have more than 10 years of experience; (c) 83% of all developers using agile methods are from small- to medium-sized firms; (d) 26% of all developers using agile methods have had improvements of 50% or greater; and (e) developers using all aspects of agile methods produced better e-commerce websites.

TABLE II
SUMMARY OF RECENT STUDIES AND SURVEYS OF AGILE METHODS

Year	Source	Findings	Responses
1998	Harvard (Thomke and Reinertsen, 1998)	50% reduction in engineering effort 55% improvement in time to market 925% improvement in number of changes allowed	391
1998	Harvard (MacCormack, 1998)	48% productivity increase over traditional methods 38% higher quality associated with more design effort 50% higher quality associated with iterative development	29
1999	Boston College (Fichman and Moses, 1999)	38% reduction in time to produce working software 50% time to market improvement 50% more capabilities delivered to customers	28
2003	Reifer Consultants (Reifer, 2003)	20% reported productivity gains 10% reported cost reductions 53% reported time-to-market improvements	78
2003	Shine Technologies (Johnson, 2003)	49% experienced cost reductions 93% experienced productivity increases 88% experienced customer satisfaction improvements	131
2004	CIO Magazine (Prewitt, 2004)	28% had been using agile methods since 2001 85% initiated enterprise-wide agile methods initiatives 43% used agile methods to improve growth and marketshare	100
2006	Digital Focus (Digital Focus, 2006)	27% of software projects used agile methods 23% had enterprise-wide agile methods initiatives 51% used agile methods to speed-up development	136
2006	Version One (Version One, 2006)	86% reported time-to-market improvements 87% reported productivity improvements 92% reported ability to dynamically change priorities	722
2006	AmbySoft (Ambler, 2006)	41% of organizations used agile methods 44% reported improved productivity, quality, and costs 38% reported improvements in customer satisfaction levels	4,232
2007	AmbySoft (Ambler, 2007)	69% of organizations had adopted agile methods 89% of agile projects had a success rate of 50% or greater 99% of organizations are now using iterative development	781
2007	UMUC (Rico, 2007)	70% of developers using most aspects of agile methods 26% of developers had improvements of 50% or greater Agile methods are linked to improved website quality	250

8. Conclusions

The gaps and problem areas in the literature associated with agile methods and website quality are numerous. First, there are few scholarly studies of agile methods.

That is, this author has been unable to locate and identify very many scholarly studies containing theoretical conceptual models of agile methods. Furthermore, few of the articles in the literature review were based on systematic qualitative or quantitative studies of agile methods. The literature review only mentions textbooks and articles with notional concepts in agile methods. Most of the quantitative survey research mentioned in the literature review was of a rudimentary attitudinal nature. In addition, few of the articles mentioned in the literature review addressed all four of the factors associated with agile methods (e.g., iterative development, customer feedback, well-structured teams and flexibility). And, few of them were systematically linked to scholarly models of website quality. So, the gaps are quite clear, a dearth holistic scholarship on agile methods and scholarly outcomes such as information systems quality.

There is a clear need for new studies on agile methods. We hope to inspire the creation of a long line of scholarly studies of agile methods. Furthermore, we hope to inspire more studies that attempt to link the factors of agile methods to scholarly models of information systems quality. First and foremost, there is a need for a systematic analysis of scholarly literature associated with the factors of agile methods. Then there is a need for a scholarly theoretical model of agile methods, depicting the factors, variables and hypotheses associated with using agile methods. In addition, there is a need for an analysis of scholarly literature to identify the factors and variables associated with website quality. Finally, there is a need to identify, survey, select or develop scholarly measures and instrument items for both agile methods and information systems quality, both of which together constitute new studies of agile methods.

REFERENCES

[1] Rosen S., 1969. Electronic computers: a historical survey. *ACM Computing Surveys*, **1**(1):7–36.

[2] Denning J., 1971. Third generation computer systems. *ACM Computing Surveys*, **3**(4):175–216.

[3] Tanenbaum A., 2001. *Modern Operating Systems*. Englewood Cliffs, NJ: Prentice Hall.

[4] Carlson B., Burgess A., and Miller C., 1996. *Timeline of Computing History*. Retrieved on October 21, 2006, from http://www.computer.org/portal/cms_docs_ieeecs/ieeecs/about/history/timeline.pdf.

[5] Nerlove M., 2004. Programming languages: A short history for economists. *Journal of Economic and Social Measurement*, **29**(1–3):189–203.

[6] Sammet J. E., 1972b. Programming languages: history and future. *Communications of the ACM*, **15**(7):601–610.

[7] Pigott D., 2006. *HOPL: An Interactive Roster of Programming Languages*. Retrieved on October 21, 2006, from http://hopl.murdoch.edu.au.

[8] Chen Y., Dios R., Mili A., Wu L., and Wang K., 2005. An empirical study of programming language trends. *IEEE Software*, **22**(3):72–78.

[9] Cusumano M. A., 1991. *Japan's Software Factories: A Challenge to U.S. Management*. New York, NY: Oxford University Press.

[10] Campbell-Kelly M., 1995. Development and structure of the international software industry: 1950–1990. *Business and Economic History*, **24**(2):73–110.

[11] Steinmueller W. E., 1996. The U.S. software industry: an analysis and interpretive history. In Mowery D. C. (Ed.), *The International Computer Software Industry* (pp. 25–52). New York, NY: Oxford University Press.

[12] Johnson L., 1998. A view from the 1960s: how the software industry began. *IEEE Annals of the History of Computing*, **20**(1):36–42.

[13] Campbell-Kelly M., 2001. Not only microsoft: the maturing of the personal computer software industry: 1982–1995. *Business History Review*, **75**(1):103–146.

[14] Middleton R., and Wardley P., 1990. Information technology in economic and social history: the computer as philosopher's stone or pandora's box? *Economic History Review*, **43**(4):667–696.

[15] U.S. Department of Commerce. 2003. *Digital Economy*. Washington, DC: Author.

[16] Borck J., and Knorr E., 2005. A field guide to hosted apps. *Infoworld*, **27**(16):38–44.

[17] Reid R. H., 1997. *Architects of the Web: 1,000 Days that Build the Future of Business*. New York, NY: John Wiley and Sons.

[18] Leiner B., Cerf V. G., Clark D. D., Kahn R. E., Kleinrock L., Lynch D. C., et al. 1997. The past and future history of the internet. *Communications of the ACM*, **40**(2):102–108.

[19] Mowery D. C., and Simcoe T., 2002. Is the internet a US invention? An economic and technological history of computer networking. *Research Policy*, **31**(8/9):1369–1387.

[20] Kalakota R., and Whinston A., 1996. *Electronic Commerce: A Manager's Guide*. Reading, MA: Addison Wesley.

[21] U.S. Census Bureau. 2006. *E-stats*. Washington, DC: Author.

[22] Mandell L., 1977. Diffusion of EFTS among national banks. *Journal of Money, Credit, and Banking*, **9**(2):341–348.

[23] Anonymous. 1965. A fascinating teller. *Banking*, **58**(3):95–95.

[24] Ellis G. H., 1967. The fed's paperless payments mechanism. *Banking*, **67**(60):100–100.

[25] New York Stock Exchange 2006. *NYSE Timeline of Technology*. Retrieved on October 23, 2006, from http://www.nyse.com/about/history/timeline_technology.html.

[26] Anonymous. 1971. Tela-fax takes dead aim on checks, credit cards. *Banking*, **64**(3):52–53.

[27] Anonymous. 1975. Bank patents its EFT system. *Banking*, **67**(1):88–88.

[28] Lynch J. E., 1990. The impact of electronic point of sale technology (EPOS) on marketing strategy and retailer-supplier relationships. *Journal of Marketing Management*, **6**(2):157–168.

[29] Accredited Standards Committee. 2006. *The Creation of ASC X12*. Retrieved on October 22, 2006, from http://www.x12.org/x12org/about/X12History.cfm.

[30] Smith A., 1988. EDI: will banks be odd man out? *ABA Banking Journal*, **80**(11):77–79.

[31] Internet Retailer. 2007. *Internet Retailer 2007 Edition Top 500 Guide: Profiles and Statistics of America's 500 Largest Retail Web Sites Ranked by Annual Sales*. Chicago, IL: Vertical Web Media, LLC.

[32] Teorey T. J., and Fry J. P., 1980. The logical record access approach to database design. *ACM Computing Surveys*, **12**(2):179–211.

[33] American National Standards Institute. 1988. *Information Resource Dictionary System* (ANSI X3.138–1988). New York, NY: Author.

[34] Dolk D. R., and Kirsch R. A., 1987. A relational information resource dictionary system. *Communications of the ACM*, **30**(1):48–61.

[35] Lombardi L., 1961. Theory of files. *Communications of the ACM*, **4**(7):324–324.

[36] Bachman C. W., 1969. Data structure diagrams. *ACM SIGMIS Database*, **1**(2):4–10.

[37] Dodd G. G., 1969. Elements of data management systems. *ACM Computing Surveys*, **1**(2):117–133.

[38] Codd E. F., 1970. A relational model of data for large shared data banks. *Communications of the ACM*, **13**(6):377–387.

[39] Cardenas A. F., 1977. Technology for automatic generation of application programs. *MIS Quarterly*, **1**(3):49–72.

[40] Montalbano M., 1962. Tables, flowcharts, and program logic. *IBM Systems Journal*, **1**(1):51–63.

[41] Oldfather P. M., Ginsberg A. S., and Markowitz H. M., 1966. *Programming by Questionnaire: How to Construct a Program Generator* (RM-5128-PR). Santa Monica, CA: The RAND Corporation.

[42] Teichroew D., and Sayani H., 1971. Automation of system building. *Datamation*, **17**(16):25–30.

[43] Sammet J. E., 1972a. An overview of programming languages for special application areas. *Proceedings of the Spring Joint American Federation of Information Processing Societies Conference (AFIPS 1972)*, Montvale, New Jersey, USA, 299–311.

[44] Boehm B. W., and Ross R., 1988. Theory W software project management: a case study. *Proceedings of the 10th International Conference on Software Engineering, Singapore*, 30–40.

[45] Anderson R. M., 1966. Management controls for effective and profitable use of EDP resources. *Proceedings of the 21st National Conference for the Association for Computing Machinery, New York, NY, USA*, 201–207.

[46] Fisher A. C., 1968. Computer construction of project networks. *Communications of the ACM*, **11**(7):493–497.

[47] Merwin R. E., 1972. Estimating software development schedules and costs. *Proceedings of the Ninth Annual ACM IEEE Conference on Design Automation, New York, NY, USA*, 1–4.

[48] Jones T. C., 1978. Measuring programming quality and productivity. *IBM Systems Journal*, **17**(1):39–63.

[49] Ives B., and Olson M. H., 1984. User involvement and MIS success: a review of research. *Management Science*, **30**(5):586–603.

[50] Dunn O. E., 1966. Information technology: a management problem. *Proceedings of the Third ACM IEEE Conference on Design Automation, New York, NY, USA*, 5.1–5.29.

[51] Fitch A. E., 1969. A user looks at DA: yesterday, today, and tomorrow. *Proceedings of the Sixth ACM IEEE Conference on Design Automation, New York, NY, USA*, 371–382.

[52] Milne M. A., 1971. CLUSTR: a program for structuring design problems. *Proceedings of the Eighth Annual ACM IEEE Design Automation Conference*, Atlantic City, New Jersey, USA, 242–249.

[53] Miller L. A., 1974. Programming by non-programmers. *International Journal of Man-Machine Studies*, **6**(2):237–260.

[54] Bechtolsheim A., 1978. Interactive specification of structured designs. *Proceedings of the 15th Annual ACM IEEE Design Automation Conference*, Las Vegas, Nevada, USA, 261–263.

[55] Dijkstra E. W., 1969. *Notes on structured programming* (T.H.-Report 70-WSK-03). Eindhoven, Netherlands: Technological University of Eindhoven.

[56] Wirth N., 1971. Program development by stepwise refinement. *Communications of the ACM*, **14**(4):221–227.

[57] Stevens W. P., Myers G. J., and Constantine L. L., 1974. Structured design. *IBM Systems Journal*, **13**(2):115–139.

[58] Yourdon E., 1976. The emergence of structured analysis. *Computer Decisions*, **8**(4):58–59.

[59] Clarke E. M., and Wing J. M., 1996. Formal methods: state of the art and future directions. *ACM Computing Surveys*, **28**(4):626–643.

[60] Hoare C. A. R., 1969. An axiomatic basis for computer programming. *Communications of the ACM*, **12**(10):576–583.

[61] Wegner P., 1972. The vienna definition language. *ACM Computing Surveys*, **4**(1):5–63.

[62] Hoare C. A. R., 1978. Communicating sequential processes. *Communications of the ACM*, **21**(8):666–677.

[63] Linger R. C., Mills H. D., and Witt B. I., 1979. *Structured programming: Theory and Practice.* Reading, MA: Addison-Wesley.

[64] Shapiro S., 1997. Splitting the difference: the historical necessity of synthesis in software engineering. *IEEE Annals of the History of Computing,* **19**(1):20–54.

[65] Van Den Bosch F., Ellis J. R., Freeman P., Johnson L., McClure C. L., Robinson D., et al. 1982. Evaluation of software development life cycle: methodology implementation. *ACM SIGSOFT Software Engineering Notes,* **7**(1):45–60.

[66] Royce W. W., 1970. Managing the development of large software systems. *Proceedings of the Western Electronic Show and Convention (WESCON 1970),* Los Angeles, California, USA, 1–9.

[67] Basili V. R., and Turner J., 1975. Iterative enhancement: a practical technique for software development. *IEEE Transactions on Software Engineering,* **1**(4):390–396.

[68] Bauer F. L., 1976. Programming as an evolutionary process. *Proceedings of the Second International Conference on Software Engineering,* San Francisco, California, USA, 223–234.

[69] Cave W. C., and Salisbury A. B., 1978. Controlling the software life cycle: the project management task. *IEEE Transactions on Software Engineering,* **4**(4):326–337.

[70] Boehm B. W., 1986. A spiral model of software development and enhancement. *ACM SIGSOFT Software Engineering Notes,* **11**(4):14–24.

[71] Belz F. C., 1986. Applying the spiral model: observations on developing system software in ada. *Proceedings of the 4th Annual National Conference on Ada Technology,* Atlanta, Georgia, USA, 57–66.

[72] Iivari J., 1987. A hierarchical spiral model for the software process. *ACM SIGSOFT Software Engineering Notes,* **12**(1):35–37.

[73] Sauer C., Jeffery D. R., Land L., and Yetton P., 2000. The effectiveness of software development technical reviews: a behaviorally motivated program of research. *IEEE Transactions on Software Engineering,* **26**(1):1–15.

[74] Weinberg G. M., 1971. *The Psychology of Computer Programming.* New York, NY: Van Nostrand Reinhold.

[75] Mills H. D., 1971. *Chief Programmer Teams: Principles and Procedures* (IBM Rep. FSC 71–5108). Gaithersburg, MD: IBM Federal Systems Division.

[76] Waldstein N. S., 1974. *The Walk Thru: A Method of Specification Design and Review* (TR 00.2536). Poughkeepsie, NY: IBM Corporation.

[77] Fagan M. E., 1976. Design and code inspections to reduce errors in program development. *IBM Systems Journal,* **15**(3):182–211.

[78] U.S. Department of Defense. 1976. *Military Standard: Technical Reviews and Audits for Systems, Equipments, and Computer Software* (MIL-STD-1521A). Hanscom AFB, MA: Author.

[79] Rosson M. B., and Alpert S. R., 1990. The cognitive consequences of object oriented design. *Human Computer Interaction,* **5**(4):345–379.

[80] Dahl O. J., and Nygaard K., 1966. Simula: an algol based simulation language. *Communications of the ACM,* **9**(9):671–678.

[81] Kay A., 1968. *Flex: A Flexible Extensible Language.* Unpublished master's thesis, University of Utah, Salt Lake City, UT, United States.

[82] Kay A., 1969. *The Reactive Engine.* Unpublished doctoral dissertation, University of Utah, Salt Lake City, UT, United States.

[83] Kay A., 1974. *Smalltalk: A Communication Medium for Children of All Ages.* Palo Alto, CA: Xerox Palo Alto Research Center.

[84] Parnas D. L., 1972. On the criteria to be used in decomposing systems into modules. *Communications of the ACM,* **15**(12):1053–1058.

[85] Booch G., 1981. Describing software design in ada. *SIGPLAN Notices,* **16**(9):42–47.

[86] McIntyre S. C., and Higgins L. F., 1988. Object oriented systems analysis and design: methodology and application. *Journal of Management Information Systems*, 5(1):25–35.

[87] Whittaker J. A., 2000. What is software testing? And why is it so hard? *IEEE Software*, 17(1):70–79.

[88] Brown J. R., and Lipow M., 1975. Testing for software reliability. *Proceedings of the First International Conference on Reliable Software, Los Angeles, California, USA*, 518–527.

[89] Goodenough J. B., and Gerhart S. L., 1975. Toward a theory of test data selection. *Proceedings of the First International Conference on Reliable Software*, Los Angeles, California, USA, 493–510.

[90] Panzl D. J., 1976. Test procedures: a new approach to software verification. *Proceedings of the Second International Conference on Reliable Software, San Francisco, California, USA*, 477–485.

[91] Walsh D. A., 1977. Structured testing. *Datamation*, 23(7):111–111.

[92] Leblang D. B., and Chase R. P., 1984. Computer aided software engineering in a distributed environment. *Proceedings of the First ACM SIGSOFT/SIGPLAN Software Engineering Symposium on Practical Software Development Environments*, Pittsburgh, Pennsylvania, USA, 104–112.

[93] Dowson M., and Wileden J. C., 1985. Panel discussion on the software process and software environments. *Proceedings of the 8th International Conference on Software Engineering, London, England*, 302–305.

[94] Nemchinova Y., 2007. *The Feasibility of Using Software Tools in Teaching Technical Courses.* Unpublished doctoral dissertation, University of Baltimore, Baltimore, MD.

[95] Baker F. T., 1975. Structured programming in a production programming environment. *Proceedings of the First International Conference on Reliable Software*, Los Angeles, California, USA, 172–185.

[96] Bratman H., and Court T., 1975. The software factory. *IEEE Computer*, 8(5):28–37.

[97] Amey W. W., 1979. The computer assisted software engineering (CASE) system. *Proceedings of the Fourth International Conference on Software Engineering, Munich, Germany*, 111–115.

[98] Wegner P., 1980. The ada language and environment. *ACM SIGSOFT Software Engineering Notes*, 5(2):8–14.

[99] Day F. W., 1983. Computer aided software engineering (CASE). *Proceedings of the 20th Conference on Design Automation*, Miami Beach, Florida, USA, 129–136.

[100] Banker R. D., and Kauffman R. J., 1991. Reuse and productivity in integrated computer aided software engineering: an empirical study. *MIS Quarterly*, 15(3):375–401.

[101] Hsieh D., 1995. David hsieh of lbms: integrated case is dead. *VARBusiness*, 11(17):136–136.

[102] Abdel-Hamid T. K., 1988. The economics of software quality assurance: a simulation based case study. *MIS Quarterly*, 12(3):394–411.

[103] Fujii M. S., 1978. A comparison of software assurance methods. *Proceedings of the First Annual Software Quality Assurance Workshop on Functional and Performance Issues*, New York, NY, USA, 27–32.

[104] Adrion W. R., Branstad M. A., and Cherniavsky J. C., 1982. Validation, verification, and testing of computer software. *ACM Computing Surveys*, 14(2):159–192.

[105] Jones C. L., 1985. A process-integrated approach to defect prevention. *IBM Systems Journal*, 24(2):150–165.

[106] Rigby P. J., Stoddart A. G., and Norris M. T., 1990. Assuring quality in software: practical experiences in attaining ISO 9001. *British Telecommunications Engineering*, 8(4):244–249.

[107] Notkin D., 1989. The relationship between software development environments and the software process. *Proceedings of the Third ACM SIGSOFT/SIGPLAN Software Engineering Symposium on Practical Software Development Environments*, Boston, Massachusetts, USA, 107–109.

[108] Crosby P. B., 1979. *Quality is Free.* New York, NY: McGraw-Hill.

[109] Radice R. A., Harding J. T., Munnis P. E., and Phillips R. W., 1985. A programming process study. *IBM Systems Journal*, 24(2):91–101.

[110] Humphrey W. S., 1987. *Characterizing the Software Process: A Maturity Framework* (CMU/SEI-87-TR-011). Pittsburgh, PA: Software Engineering Institute.

[111] Weber C., Paulk M., Wise C., and Withey J., 1991. *Key Practices of the Capability Maturity Model* (CMU/SEI-91-TR-025). Pittsburgh, PA: Software Engineering Institute.

[112] Agarwal R., Prasad J., Tanniru M., and Lynch J., 2000. Risks of rapid application development. *Communications of the ACM*, **43**(11):177–188.

[113] Naumann J. D., and Jenkins A. M., 1982. Prototyping: the new paradigm for systems development. *MIS Quarterly*, **6**(3):29–44.

[114] Alavi M., 1985. Some thoughts on quality issues of end-user developed systems. *Proceedings of the 21st Annual Conference on Computer Personnel Research, Minneapolis, Minnesota, USA*, 200–207.

[115] Guide International, Inc. 1986. *Joint Application Design*. Chicago, IL: Author.

[116] Gane C., 1987. *Rapid Systems Development*. New York, NY: Rapid Systems Development, Inc.

[117] Martin J., 1991. *Rapid Application Development*. New York, NY: Macmillan.

[118] Frakes W. B., and Kang K., 2005. Software reuse research: status and future. *IEEE Transactions on Software Engineering*, **31**(7):529–536.

[119] McIlroy M. D., 1968. Mass produced software components. *Proceedings of the NATO Software Engineering Conference*, Garmisch, Germany, 138–155.

[120] Pyster A., 1982. Software development productivity. *Proceedings of the National ACM Conference. Dallas, Texas, USA*, 94–94.

[121] Lubars M. D., 1982. Affording higher reliability through software reusability. *ACM SIGSOFT Software Engineering Notes*, **11**(5):39–42.

[122] Zychlinski B. Z., and Palomar M. A., 1984. A software quality assurance program through reusable code. *Proceedings of the 3rd Annual International Conference on Systems Documentation*, Mexico City, Mexico, 107–113.

[123] Lim W. C., 1994. Effects of reuse on quality, productivity, and economics. *IEEE Software*, **11**(5): 23–30.

[124] Neighbors J. M., 1984. The draco approach to constructing software from reusable components. *IEEE Transactions on Software Engineering*, **10**(5):564–574.

[125] Jameson K. W., 1989. A model for the reuse of software design information. *Proceedings of the 11th International Conference on Software Engineering, Pittsburgh, Pennsylvania, USA*, 205–216.

[126] Holibaugh R., Cohen S., Kang K., and Peterson S., 1989. Reuse: where to begin and why. *Proceedings of the Conference on Tri-Ada*, Pittsburgh, Pennsylvania, USA, 266–277.

[127] D'Souza D. F., and Wills A. C., 1998. *Objects, Components, and Frameworks With UML: The Catalysis Approach*. Reading, MA: Addison Wesley.

[128] McGibbon T., 1996. *A Business Case for Software Process Improvement* (Contract Number F30602–92-C-0158). Rome, NY: Air Force Research Laboratory – Information Directorate (AFRL/IF), Data and Analysis Center for Software (DACS).

[129] Poulin J. S., 1997. *Measuring Software Reuse: Principles, Practices, and Economic Models*. Reading, MA: Addison Wesley.

[130] Lim W. C., 1998. *Managing Software Reuse: A Comprehensive Guide to Strategically Reengineering the Organization for Reusable Components*. Upper Saddle River, NJ: Prentice Hall.

[131] Edwards S. H., 1999. The state of reuse: perceptions of the reuse community. *ACM SIGSOFT Software Engineering Notes*, **24**(3):32–36.

[132] Sherif K., and Vinze A., 1999. A qualitative model for barriers to software reuse adoption. *Proceeding of the 20th International Conference on Information Systems, Charlotte, North Carolina, USA*, 47–64.

[133] Kruchten P., Obbink H., and Stafford J., 2006. The past, present, and future of software architecture. *IEEE Software*, **23**(2):22–30.

[134] Prieto-Diaz R., 1987. Domain analysis for reusability. *Proceedings of the 11th Annual International Computer Software and Applications Conference (COMPSAC 1987)*, Tokyo, Japan, 23–29.

[135] Arango G., 1988. *Domain Engineering for Software Reuse* (ICS-RTP-88–27). Irvine, CA: University of California Irvine, Department of Information and Computer Science.

[136] Horowitz B. B., 1991. *The Importance of Architecture in DoD Software* (Technical Report M91-35). Bedford, MA: The Mitre Corporation.

[137] Wegner P., Scherlis W., Purtilo J., Luckham D., and Johnson R., 1992. Object oriented megaprogramming. *Proceedings on Object Oriented Programming Systems, Languages, and Applications, Vancouver, British Columbia, Canada*, 392–396.

[138] Northrop L. M., 2002. SEI's software product line tenets. *IEEE Software*, **19**(4):32–40.

[139] Highsmith J. A., 2002. *Agile Software Development Ecosystems*. Boston, MA: Addison Wesley.

[140] Beck K., 1999. Embracing change with extreme programming. *IEEE Computer*, **32**(10):70–77.

[141] Agile Manifesto. 2001. *Manifesto for Agile Software Development*. Retrieved on November 29, 2006, from http://www.agilemanifesto.org.

[142] Millington D., and Stapleton J., 1995. Developing a RAD standard. *IEEE Software*, **12**(5):54–56.

[143] Schwaber K., 1995. Scrum development process. *Proceedings of the 10th Annual ACM Conference on Object Oriented Programming Systems, Languages, and Applications (OOPSLA 1995)*, Austin, Texas, USA, 117–134.

[144] Anderson A., Beattie R., Beck K., Bryant D., DeArment M., Fowler M., et al. 1998. Chrysler goes to extremes. *Distributed Computing Magazine*, **1**(10):24–28.

[145] O'Reilly T., 1999. Lessons from open source software development. *Communications of the ACM*, **42**(4):32–37.

[146] Cockburn A., 2002a. *Agile Software Development*. Boston, MA: Addison Wesley.

[147] Palmer S. R., and Felsing J. M., 2002. *A Practical Guide to Feature Driven Development*. Upper Saddle River, NJ: Prentice Hall.

[148] Kruchten P., 2000. *The Rational Unified Process: An Introduction*. Reading, MA: Addison Wesley.

[149] Highsmith J. A., 2000. *Adaptive Software Development: A Collaborative Approach to Managing Complex Systems*. New York, NY: Dorset House.

[150] Poppendieck M., and Poppendieck T., 2003. *Lean Software Development: An Agile Toolkit for Software Development Managers*. Boston, MA: Addison Wesley.

[151] McIlroy M. D., 1960. Macro instruction extensions of compiler languages. *Communications of the ACM*, **3**(4):214–220.

[152] Conte S. D., Dunsmore H. E., and Shen V. Y., 1986. *Software Engineering Metrics and Models*. Menlo Park, CA: Benjamin Cummings.

[153] Martin J., 1965. *Programming Real-Time Computer Systems*. Englewood Cliffs, NJ: Prentice Hall.

[154] Halstead M. H., 1977. *Elements of Software Science*. New York, NY: Elsevier North Holland.

[155] Basili V. R., and Reiter R. W., 1979. An investigation of human factors in software development. *IEEE Computer*, **12**(12):21–38.

[156] Albrecht A. J., 1979. Measuring application development productivity. *Proceedings of the IBM Applications Development Joint SHARE/GUIDE Symposium, Monterrey, California, USA*, 83–92.

[157] Kan S. H., 1995. *Metrics and Models in Software Quality Engineering*. Reading, MA: Addison-Wesley.

[158] Weinberg G. M., and Gressett G. L., 1963. An experiment in automatic verification of programs. *Communications of the ACM*, **6**(10):610–613.

[159] Youngs E. A., 1970. *Error Proneness in Programming*. Unpublished doctoral dissertation. University of North Carolina at Chapel Hill, Chapel Hill, NC, United States.

[160] Shooman M. L., and Bolsky M. I., 1975. Types, distribution, and test and correction times for programming errors. *Proceedings of the International Conference on Reliable Software*, Los Angeles, California, USA, 347–357.

[161] Lipow M., 1982. Number of faults per line of code. *IEEE Transactions on Software Engineering*, **8**(4):437–439.

[162] Shen V. Y., Yu T. J., Thebaut S. M., and Paulsen L. R., 1985. Identifying error prone software: An empirical study. *IEEE Transactions on Software Engineering*, **11**(4):317–324.

[163] Institute of Electrical and Electronics Engineers. 1990. *IEEE Standard Glossary of Software Engineering Terminology* (IEEE Std 610.12–1990). New York, NY: Author.

[164] Rubey R. J., and Hartwick R. D., 1968. Quantitative measurement of program quality. *Proceedings of the 23rd ACM National Conference*, Washington, DC, USA, 671–677.

[165] Boehm B. W., Brown J. R., Kaspar H., and Lipow M., 1973. *Characteristics of Software Quality* (TRW-SS-73-09). Redondo Beach, CA: TRW Corporation.

[166] Swanson E. B., 1976. The dimensions of maintenance. *Proceedings of the Second International Conference on Software Engineering*, San Francisco, California, USA, 492–497.

[167] Gilb T., 1977. *Software Metrics*. Cambridge, MA: Winthrop Publishers.

[168] Cavano J. P., and McCall J. A., 1978. A framework for the measurement of software quality. *Proceedings of the Software Quality Assurance Workshop on Functional and Performance Issues*, San Diego, California, USA, 133–139.

[169] Dzida W., Herda S., and Itzfeldt W. D., 1978. User perceived quality of interactive systems. *Proceedings of the Third International Conference on Software Engineering, Atlanta, Georgia, USA*, 188–195.

[170] Gaffney J. E., 1981. Metrics in software quality assurance. *Proceedings of the ACM SIGMETRICS Workshop/Symposium on Measurement and Evaluation of Software Quality*, Las Vegas, Nevada, USA, 126–130.

[171] Sunazuka T., Azuma M., and Yamagishi N., 1985. Software quality assessment technology. *Proceedings of the Eighth International Conference on Software Engineering*, London, England, 142–148.

[172] Arthur L. J., 1985. *Measuring Programmer Productivity and Software Quality*. New York, NY: John Wiley and Sons.

[173] Grady R. B., and Caswell R. B., 1987. *Software Metrics: Establishing a Company Wide Program*. Englewood Cliffs, NJ: Prentice Hall.

[174] Akiyama F., 1971. An example of software system debugging. *Proceedings of the International Federation for Information Processing Congress*, Ljubljana, Yugoslavia, 353–379.

[175] Motley R. W., and Brooks W. D., 1977. *Statistical Prediction of Programming Errors* (RADC-TR-77-175). Griffis AFB, NY: Rome Air Development Center.

[176] Potier D., Albin J. L., Ferreol R., and Bilodeau A., 1982. Experiments with computer software complexity and reliability. *Proceedings of the Sixth International Conference on Software Engineering*, Tokyo, Japan, 94–103.

[177] Halstead M. H., 1972. Natural laws controlling algorithm structure? *ACM SIGPLAN Notices*, **7**(2):19–26.

[178] Weissman L., 1973. Psychological complexity of computer programs. *ACM SIGPLAN Notices*, **8**(6):92–95.

[179] Elshoff J. L., 1976. An analysis of some commercial PL/1 programs. *IEEE Transactions on Software Engineering*, **2**(2):113–120.

[180] Dunsmore H. E., and Gannon J. D., 1979. Data referencing: an empirical investigation. *IEEE Computer*, **12**(12):50–59.

[181] Henry S., and Kafura D., 1981, Software structure metrics based on information flow. *IEEE Transactions on Software Engineering*, **7**(5):510–518.

[182] McCabe T. J., 1976. A complexity measure. *IEEE Transactions on Software Engineering*, **2**(4): 308–320.

[183] Schneidewind N. F., and Hoffmann H., 1979. An experiment in software error data collection and analysis. *IEEE Transactions on Software Engineering*, **5**(3):276–286.

[184] Woodward M. R., Hennell M. A., and Hedley D., 1979. A measure of control flow complexity in program text. *IEEE Transactions on Software Engineering*, **5**(1):45–50.

[185] Dunsmore H. E., and Gannon J. D., 1980. Analysis of the effects of programming factors on programming effort. *Journal of Systems and Software*, **1**(2):141–153.

[186] Shooman M. L., 1983. *Software Engineering*. New York, NY: McGraw Hill.

[187] Zolnowski J. C., and Simmons D. B., 1981. Taking the measure of program complexity. *Proceedings of the AFIPS National Computer Conference*, Chicago, Illinois, USA, 329–336.

[188] Myers G. J., 1977. An extension to the cyclomatic measure of program complexity. *SIGPLAN Notices*, **12**(10):61–64.

[189] Hansen W. J., 1978. Measurement of program complexity by the pair (cyclomatic number, operator count). *ACM SIGPLAN Notices*, **13**(3):29–33.

[190] Oviedo E. I., 1980. Control flow, data flow, and program complexity. *Proceedings of the Fourth International IEEE Computer Software and Applications Conference (COMPSAC 1980)*, Chicago, Illinois, USA, 146–152.

[191] Card D. N., and Glass R. L., 1990. *Measuring Software Design Quality*. Englewood Cliffs, NJ: Prentice Hall.

[192] Lo B., 1992. *Syntactical Construct Based APAR Projection* (Technical Report). San Jose, CA: IBM Santa Teresa Research Laboratory.

[193] Myers G. J., 1976. *Software Reliability: Principles and Practices*. New York, NY: John Wiley and Sons.

[194] Musa J. D., 1999. *Software Reliability Engineering*. New York, NY: McGraw Hill.

[195] Pham H., 2000. *Software Reliability*. Singapore: Springer Verlag.

[196] Jelinski Z., and Moranda P. B., 1972. Software reliability research. In Freiberger W. (Ed.), *Statistical Computer Performance Evaluation* (pp. 465–484). New York, NY: Academic Press.

[197] Schick G. J., and Wolverton R. W., 1978. An analysis of competing software reliability analysis models. *IEEE Transactions on Software Engineering*, **4**(2):104–120.

[198] Moranda P. B., 1979. An error detection model for application during software development. *IEEE Transactions on Reliability*, **28**(5):325–329.

[199] Goel A. L., and Okumoto K., 1979. Time dependent error detection rate model for software and other performance measures. *IEEE Transactions on Reliability*, **28**(3):206–211.

[200] Littlewood B., 1979. Software reliability model for modular program structure. *IEEE Transactions on Reliability*, **28**(3):241–246.

[201] Sukert A. N., 1979. Empirical validation of three software error prediction models. *IEEE Transactions on Reliability*, **28**(3):199–205.

[202] Coutinho J. S., 1973. Software reliability growth. *Proceedings of the IEEE Symposium on Computer Software Reliability*, New York, NY, USA, 58–64.

[203] Wall J. K., and Ferguson P. A., 1977. Pragmatic software reliability prediction. *Proceedings of the Annual Reliability and Maintainability Symposium*, Piscataway, New Jersey, USA, 485–488.

[204] Huang X. Z., 1984. The hypergeometric distribution model for predicting the reliability of software. *Microelectronics and Reliability*, **24**(1):11–20.

[205] Musa J. D., Iannino A., and Okumoto K., 1987. *Software Reliability: Measurement, Prediction, and Application*. New York, NY: McGraw Hill.

[206] Ohba M., 1984. Software reliability analysis models. *IBM Journal of Research and Development*, **21**(4):428–443.

[207] Yamada S., Ohba M., and Osaki S., 1983. S shaped reliability growth modeling for software error prediction. *IEEE Transactions on Reliability*, **32**(5):475–478.

[208] Thayer C. H., 1958. Automation and the problems of management. *Vital Speeches of the Day*, **25**(4):121–125.

[209] Hardin K., 1960. Computer automation, work environment, and employee satisfaction: a case study. *Industrial and Labor Relations Review*, **13**(4):559–567.

[210] Kaufman S., 1966. The IBM information retrieval center (ITIRC): system techniques and applications. *Proceedings of the 21st National Conference for the Association for Computing Machinery*, New York, NY, USA, 505–512.

[211] Lucas H. C., 1973. User reactions and the management of information services. *Management Informatics*, **2**(4):165–162.

[212] Bailey J. E., and Pearson S. W., 1983. Development of a tool for measuring and analyzing computer user satisfaction. *Management Science*, **29**(5):530–545.

[213] Lucas H. C., 1974. Measuring employee reactions to computer operations. *Sloan Management Review*, **15**(3):59–67.

[214] Maish A. M., 1979. A user's behavior toward his MIS. *MIS Quarterly*, **3**(1):39–52.

[215] Lyons M. L., 1980. Measuring user satisfaction: the semantic differential technique. *Proceedings of the 17th Annual Conference on Computer Personnel Research*, Miami, Florida, USA, 79–87.

[216] Pearson S. W., and Bailey J. E., 1980. Measurement of computer user satisfaction. *ACM SIGMETRICS Performance Evaluation Review*, **9**(1):9–68.

[217] Walsh M. D., 1982. Evaluating user satisfaction. *Proceedings of the 10th Annual ACM SIGUCCS Conference on User Services*, Chicago, Illinois, USA, 87–95.

[218] Ives B., Olson M. H., and Baroudi J. J., 1983. The measurement of user information satisfaction. *Communications of the ACM*, **26**(10):785–793.

[219] Joshi K., Perkins W. C., and Bostrom R. P., 1986. Some new factors influencing user information satisfaction: implications for systems professionals. *Proceedings of the 22nd Annual Computer Personnel Research Conference*, Calgary, Canada, 27–42.

[220] Baroudi J. J., and Orlikowski W. J., 1988. A short form measure of user information satisfaction: a psychometric evaluation and notes on use. *Journal of Management Information Systems*, **4**(4):44–59.

[221] Doll W. J., and Torkzadeh G., 1988. The measurement of end user computing satisfaction. *MIS Quarterly*, **12**(2):258–274.

[222] Kekre S., Krishnan M. S., and Srinivasan K., 1995. Drivers of customer satisfaction in software products: implications for design and service support. *Management Science*, **41**(9):1456–1470.

[223] Lindroos K., 1997. Use quality and the world wide web. *Information and Software Technology*, **39**(12):827–836.

[224] Dreze X., and Zufryden F., 1997. Testing web site design and promotional content. *Journal of Advertising Research*, **37**(2):77–91.

[225] Selz D., and Schubert P., 1997. Web assessment: a model for the evaluation and the assessment of successful electronic commerce applications. *Electronic Markets*, **7**(3):46–48.

[226] Chen Q., and Wells W. D., 1999. Attitude toward the site. *Journal of Advertising Research*, **39**(5): 27–37.

[227] Szymanski D. M., and Hise R. T., 2000. E-satisfaction: an initial examination. *Journal of Retailing*, **76**(3):309–322.

[228] Barnes S. J., and Vidgen R. T., 2000. Webqual: an exploration of web site quality. *Proceedings of the Eighth European Conference on Information Systems, Vienna, Austria*, 298–305.

[229] Barnes S. J., and Vidgen R. T., 2001. An evaluation of cyber bookshops: the webqual method. *International Journal of Electronic Commerce*, **6**(1):11–30.

[230] Cho N., and Park S., 2001. Development of electronic commerce user consumer satisfaction index (ECUSI) for internet shopping. *Industrial Management and Data Systems*, **101**(8/9):400–405.

[231] Yoo B., and Donthu N., 2001. Developing a scale to measure the perceived quality of an internet shopping site (sitequal). *Quarterly Journal of Electronic Commerce*, **2**(1):31–45.

[232] Janda S., Trocchia P. J., and Gwinner K. P., 2002. Consumer perceptions of internet retail service quality. *International Journal of Service Industry Management*, **13**(5):412–433.

[233] McKinney V., Yoon K., and Zahedi F., 2002. The measurement of web customer satisfaction: an expectation and disconfirmation approach. *Information Systems Research*, **13**(3):296–315.

[234] Wolfinbarger M., and Gilly M. C., 2003. Etailq: dimensionalizing, measuring, and predicting etail quality. *Journal of Retailing*, **79**(3):183–198.

[235] Takeuchi H., and Nonaka I., 1986. The new product development game. *Harvard Business Review*, **64**(1):137–146.

[236] Sulack R. A., Lindner R. J., and Dietz D. N., 1989. A new development rhythm for AS/400 software. *IBM Systems Journal*, **28**(3):386–406.

[237] Cusumano M. A., and Selby R. W., 1995. *Microsoft Secrets: How the World's Most Powerful Software Company Creates Technology, Shapes Markets, and Manages People*. New York, NY: The Free Press.

[238] Cusumano M. A., and Yoffie D. B., 1998. *Competing on Internet Time: Lessons from Netscape and Its Battle with Microsoft*. New York, NY: The Free Press.

[239] MacCormack A., 1998. *Managing Adaptation: An Empirical Study of Product Development in Rapidly Changing Environments*. Unpublished doctoral dissertation, Harvard University, Boston, MA, United States.

[240] Thomke S., and Reinertsen D., 1998. Agile product development: managing development flexibility in uncertain environments. *California Management Review*, **41**(1):8–30.

[241] Fichman R. G., and Moses S. A., 1999. An incremental process for software implementation. *Sloan Management Review*, **40**(2):39–52.

[242] Reifer D. J., 2003. The business case for agile methods/extreme programming (XP). *Proceedings of the Seventh Annual PSM Users Group Conference*, Keystone, Colorado, USA, 1–30.

[243] Johnson M., 2003. *Agile Methodologies: Survey Results*. Victoria, Australia: Shine Technologies.

[244] Prewitt E., 2004. The agile 100. *CIO Magazine*, **17**(21):4–7.

[245] Digital Focus. 2006. *Agile 2006 Survey: Results and Analysis*. Herndon, VA: Author.

[246] Version One. 2006. *The State of Agile Development*. Apharetta, GA: Author.

[247] Ambler S. W., 2006. *Agile Adoption Rate Survey: March 2006*. Retrieved on September 17, 2006, from http://www.ambysoft.com/downloads/surveys/AgileAdoptionRates.ppt.

[248] Ambler S. W., 2007. *Agile Adoption Survey: March 2007*. Retrieved on July 23, 2007, from http://www.ambysoft.com/downloads/surveys/AgileAdoption2007.ppt.

[249] Rico D. F., 2007. *Effects of Agile Methods on Website Quality for Electronic Commerce*. Unpublished doctoral dissertation, University of Maryland University College, Adelphi, MD, United States.

[250] Rico D. F., Sayani H. H., Stewart J. J., and Field R. F., 2007. A model for measuring agile methods and website quality. *TickIT International*, **9**(3):3–15.

[251] Rico D. F., 2007. Effects of agile methods on electronic commerce: Do they improve website quality? *Proceedings of the 40th Annual Hawaii International Conference on System Sciences (HICSS 2007)*, Waikaloa, Big Island, Hawaii.

[252] Rico D. F., 2008. Effects of agile methods website quality for electronic commerce. *Proceedings of the 41st Annual Hawaii International Conference on System Sciences (HICSS 2008)*, Waikaloa, Big Island, Hawaii.

Testing with Software Designs

ALIREZA MAHDIAN AND ANNELIESE A. ANDREWS

Department of Computer Science
University of Denver
Denver, CO 80208, USA
contact: andrews@cs.du.edu, 303-871-3374

Abstract

This chapter explores current state-of-the-art techniques that have been used in software design testing, either to test the designs or to test implementations against the designs. A common design notation that is in use today is UML. Originally, techniques have been designed with the intentions of testing implementations against their design artifacts provided in UML, but there are also techniques that test the design artifacts directly. Given that UML is a defacto standard in design notations, this chapter mostly focuses on testing techniques that either test designs in UML directly, or use a design in UML to test its implementation. As appropriate, we refer to general testing principles and techniques to illustrate their application in the context of testing with software designs. The chapter covers relevant testing criteria, testing techniques based on the type of UML diagram and automated test generation.

1. Introduction

Building of quality software is a major concern for software development organizations. Effective testing is an important part of a quality development effort. Software testing research has provided a wide array of test and test generation methods, particularly for code. More recently, testing methods have been proposed for designs. Given that the Unified Modelling Language (UML) [35] is the defacto standard for design artifacts, most of the testing approaches revolve around designs using various UML diagrams. UML has made it possible to describe designs with a uniform notation at a variety of design levels ranging from conceptual to detailed design [8].

There are two basic ways to use UML design artifacts for testing; either for testing an implementation against its design (e.g., [10]), or to test the designs themselves to evaluate their quality (e.g., [33]). This chapter covers both uses of UML artifacts for testing. Testing rather than the more traditional inspection of design artifacts becomes important for the following reasons:

- The complexity and sheer size of many designs.
- UML designs contain multiple notations allowing for complex interactions between design artifacts.

Complex systems such as telecommunication systems can lead to hundreds of pages of UML design in various notations, from Class Diagrams, to Sequence Diagrams, to State Charts, etc. These diagrams commonly interact in ways such that their correctness is far from being easily determined. Given the complexity and multiple views through multiple models such as Class Diagrams, Sequence Diagrams, constraints in Object Control Language (OCL) [35] and the like, the evaluation of the designs can be difficult.

This results in an urgent need for devising a systematic (testing) approach to design evaluation. To be comprehensive, one must provide:

- test criteria that cover all elements of UML models,
- test-generation techniques and tools to automate them
- test-execution environments including coverage measurement,
- test-validation techniques and tools (oracles).

This chapter surveys the state-of-the art techniques with respect to testing with UML designs (*with UML*) and testing UML designs themselves (*for UML*). As can be expected, not all areas of systematic testing listed above have been covered widely. Currently, a comprehensive set of testing criteria, approaches and tools does not exist. This chapter analyses the currently available techniques. Roughly, more testing techniques have been proposed for more commonly occurring diagrams, such as Class Diagrams, Collaboration or sequence Diagrams, or Statecharts. There are more

TABLE I

CLASSIFICATION OF EXISTING WORK ON TESTING (WITH) SOFTWARE DESIGNS

Design Methods	With UML	For UML
Testing criteria	(Abdurazik and Offut, 2000, [1])	(Ghosh et al., 2003, [15])
		(Andrews et al., 2000, [2])
UML artifacts:	(Briand and Labiche, 2001, [10])	(Pilskalns et al., 2007, [33])
Multiple diagrams		(Pilskalns et al., 2003, [34])
		(Trong et al., 2005, [38])
Statechart	(Briand et al., 2003, [9])	
	(Gnesi et al., 2004, [16])	
	(Latella and Massink, 2001, [24])	
	(Offut and Abdurazik, 1999, [29])	
Class diagram	(Briand and Labiche, 2001, [10])	(Gogolla et al., 2003, [17])
	(Scheetz et al., 1999, [36])	(Pilskalns et al., 2007, [33])
	(von Mayrhauser et al., 2000, [39])	(Pilskalns et al., 2003, [34])
		(Trong et al., 2005, [38])
Collaboration/sequence	(Briand and Labiche, 2001, [10])	(Pilskalns et al., 2007, [33])
diagram		(Pilskalns et al., 2003, [34])
		(Trong et al., 2005, [38])
Automated test	(Briand and Labiche, 2001, [10])	(Pilskalns et al., 2007, [33])
generation	(von Mayrhauser et al., 2000, [39])	(Knight, 2005, [23])

proposed techniques than automated test generation tools. Fewer solutions exist for regression testing. Table I shows existing testing approaches, classified based on whether they are *with UML* or *for UML*. In addition, the table distinguishes between the areas of testing listed above (rows in Table I) as well as whether a particular technique targets a specific type of UML notation or covers multiple types of diagrams. Section 2 describes test adequacy criteria related to testing UML. Section 3 surveys testing methods, *with UML* and *for UML*. Section 4 discusses the limits of current state-of-the-art techniques and suggests future testing improvements.

2. Testing Criteria

A test adequacy criterion defines requirements for sufficient testing. This area of testing has been extensively analysed for white-box testing of code. For example, a common white-box adequacy criterion is branch adequacy. If a program P is represented by a flowchart, then a branch is an edge of the flowchart. A test set T is branch adequate for P, provided for every branch b of P, there is some t in T which causes b to be traversed.

Associated with the definition of test criteria is the question of whether they are 'reasonable'. This question can be assessed analytically or empirically. An analytic approach in [41] defines a general axiomatic theory for test adequacy criteria. The motivation behind this work is to *understand the strengths and weaknesses of the proposed adequacy criteria and guide the definition of new adequacy criteria.*

Test adequacy criteria were originally defined for code, either for white box testing [40, 41] or for testing code against its specification [30]. Test adequacy criteria for UML designs have been developed more recently [2].

Summarizing [41]:

- The first four properties state that:
    1. Every program must be testable.
    2. An adequacy criterion must be satisfiable in a non-trivial way.
    3. A program which has not been tested at all should not be deemed adequately tested.
    4. Once a program has been adequately tested, no amount of additional test data can result in an inadequately tested program.

- The next three properties state that:
    1. Programs which are closely related either syntactically or semantically, but not both, may well require different test data.
    2. The fact that all parts of a program have been adequately tested does not necessarily imply that the entire program has been adequately tested.
    3. Even though a program has been adequately tested, it does not follow that each of its components has been adequately tested. This is due to the fact that programs may contain unreachable code.

Weyuker [40] extended her previous work by showing that even though the properties were useful in assessing the strength and weaknesses of the proposed program-based adequacy criteria, they could all be simultaneously satisfied by obviously unsuitable adequacy criteria. Weyuker then adds three new properties to substantially strengthen the set and, in particular, to rule out unsuitable adequacy criteria. The three additional properties are:

- *Renaming property:* renaming of identifiers does not change an adequate test set into an inadequate one.
- *Complexity property:* a program exists for every minimal test suite size n.
- *Statement coverage property:* test criteria must force statement coverage.

With some modifications, these properties can be extended to cover design-based criteria:

- **Property 1**: For every design, there exists an adequate test set. A design can be implemented in different ways, and there exists an adequate test set for each of those implementations. If each implementation can be tested adequately by a particular test set, then there has to exist a test set that can adequately test the set of all these implementations and thus the design itself.

- **Property 2**: There is a design and a corresponding test set that can non-exhaustively and adequately, test that design. This means that a criterion cannot always require an exhaustive test set for every possible design.

- **Property 3**: If there exists a subset of a test set adequate for a design, then that test set is also adequate for that particular design.

- **Property 4**: An empty set is not adequate for any design. This means that a design always needs to be tested.

- **Property 5**: There are designs that are equivalent but require different adequate test sets. Since a functionality can be accomplished by different designs, we can have equivalent designs. The fact that two designs are equivalent does not imply that they necessarily have the same adequate test set.

- **Property 6**: If two designs are similar in structure, this does not necessarily mean that they require the same adequate test set.

- **Property 7**: If a design is composed of smaller components, then an adequate test set for the whole design does not guarantee an adequate test set for each component of the design as an independent individual component.

- **Property 8**: If a design is composed of two or more components, then an adequate test set would be more than just the union of adequate test sets for each design component. This is due to the complexity added to the whole design because of added interactions among the different components in the design.

While Weyuker [40, 41] did not use any concepts from UML, they influenced the definition of test adequacy criteria for UML designs [2, 15]. In [15] and [2], a set of testing criteria for UML-based design artifacts has been presented. The method incorporates the use of test adequacy criteria based on UML model elements in class diagrams and interaction diagrams. Class diagram criteria are used to determine the object configurations on which tests are run, while interaction diagram criteria are used to determine the sequences of messages that should be tested. Table II summarizes the criteria derived for the class and collaboration diagrams. The basic approach for defining these testing criteria is to define key building blocks for each type of diagram (class and collaboration diagrams) and to enforce block coverage as a test requirement. For class diagrams, these blocks are:

- Generalization relationships.
- Association-end multiplicities.
- Class attributes.

TABLE II
TEST CRITERIA FOR CLASS AND COLLABORATION DIAGRAMS ADAPTED FROM [2]

Association-end multiplicity (AEM) criterion
Given a test set T and a system model SM, T must cause each representative multiplicity-pair in SM to be created.

Generalization (GN) criterion
Given a test set T and a system model SM, T must cause every specialization defined in a generalization relationship to be created.

Class attribute (CA) criterion
Given a test set T, a system model SM, and a class C, T must cause a set of representative attribute value combinations in each instance of class C to be created.

Condition coverage (Cond) criterion
Given a test set T and a collaboration diagram CD, T must cause each condition in each decision for evaluation of both TRUE and FALSE.

Full predicate coverage (FP) criterion
Given a test set T and a collaboration diagram CD, T must cause each clause in every condition in CD to take the values of TRUE and FALSE, while all other clauses in the predicate (condition) have values such that the value of the predicate will always be the same as the clause being tested.

Each message on link (EML) criterion
Given a test set T, a collaboration diagram CD, T must cause each message on a link connecting two objects in CD to be executed at least once.

All message paths (AMP) criterion
Given a test set T, a collaboration diagram CD, T must cause each possible message path (sequence of message numbers) in CD to be taken at least once.

Collection coverage (Coll) criterion
Given a test set T, a collaboration diagram CD, T must test each interaction with collection objects of various representative sizes at least once.

For collaboration diagrams, they are:

• Conditions in the collaboration diagram.
• Each clause within each condition.
• Each message.
• Each message path.

The AEM and CA criteria are expressed in terms of representative values. A form of category-partition testing adapted to UML diagrams can be used to establish the set of representative values. The value domain is partitioned into equivalence classes, and one value from each class is selected for the representative values.

In [2], preliminary results of a case study are presented. They are based on a fault model for various types of UML design artifacts. Faults are classified as incorrect sequence numbering, missing flows, and dataflow gaps. It is shown that these types of faults can be uncovered by the proposed criteria.

Offut et al. [29] propose test criteria for UML statecharts. They defined four testing criteria:

1. *Transition coverage: every transition in the statechart must be traversed.*
2. *Full predicate coverage: the test set should include a pair of tests for each clause c in each predicate P so that the value of P directly correlates with the value of c.*
3. *Transition pair coverage: for each pair of adjacent transitions $S_i : S_j$ and $S_j : S_k$ in SG, T contains a test that traverses the pair of transitions in sequence.*
4. *Complete sequence coverage: meaningful sequences of transitions should be defined for the statechart.*

Collaboration diagrams consist of messages that are passed between objects and their sequences, thus they provide design-level control and data flow information. Because many testing techniques use data flow and control flow information, collaboration diagrams play an important role in testing designs. Abdurazik et al. [1] defined testing criteria specifically for collaboration diagrams. They define testing criteria for both static and dynamic testing of UML collaboration diagrams. The static testing focuses on checking of the code without executing it as opposed to dynamic checking where the software is executed on some inputs. The items that should be used in static testing are described as follows:

- *Classifier roles:* An object plays a classifier role in a collaboration diagram. Those classifier roles that originate from the same class should be tested to see if they have all the required attributes and operations.

- *Collaborating pairs:* Any pair of objects that are connected to each other via a link in the collaboration diagram are collaborating pairs. Each collaborating pair should be tested at least once.

- *Message or stimulus:* Testing a message will reveal most integration problems. Return value type, thread of control and input parameters are some aspects of a message that needs to be tested. A stimulus is an instance of a message.

- *Local variable definition-usage link pairs:* Variable definition-usage link pair is a pair of link which consists of the message that defines the variable and the first message that uses the variable. Testing of variable definition-usage link pair includes traversing the links between these two. This test would help the tester in finding data flow anomalies.

For dynamic testing, the test set should have at least one test case per collaboration diagram that executes the complete message sequence of that collaboration diagram. To check that the system will produce an event trace which conforms to the message sequence path of the collaboration diagrams, *instrumentations* can be inserted at the entry point of each method in the message path sequence in the original program. These *instruments* are more like watchdogs that help keep track of run-time interaction traces.

3. Design Evaluation Methods

In this section, we review testing methods that use UML. The first three methods use UML diagrams as input artifacts in order to test code (**With UML**), as opposed to other methods which use UML diagrams to test the UML design itself (**For UML**). We will illustrate each testing method with the same example throughout the chapter. The example represents a simplified course registration system. The registrar is the only actor who interacts with the system. Figure 1 represents the class diagram for the registration system.

The main functions of the system are classified into two groups based on whether they are initiated by the registrar or executed automatically by the system itself. The following functions are initiated by the registrar:

- Add /remove student.
- Add /remove instructor.
- Add /remove department.
- Add /remove course and sections.
- Add /remove course catalog.
- Add course to student's schedule.
- Display course list, student list, department list and course catalog.

The following functions are executed automatically based on the system time:

- Archiving the registration information for the current semester at the end of the semester.
- Removing classes that do not have the minimum number of students by the end of registration deadline.
- Finalizing student schedules and list of students in each course right after the deadline for dropping a course is over.

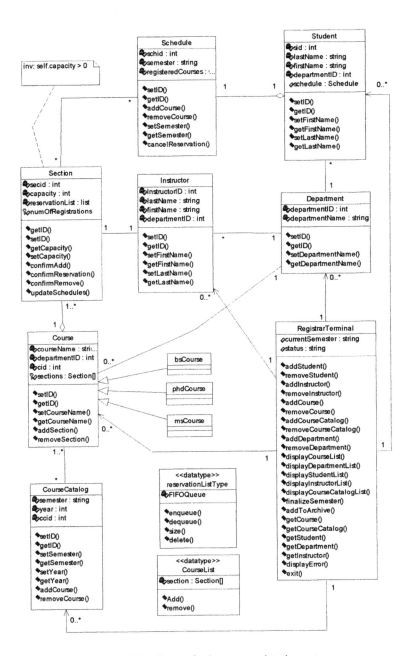

FIG. 1. Class diagram for the course registration system.

3.1   With UML

3.1.1   TOTEM

Briand et al. [10] introduce an approach to derive system test requirements. With the availability of detailed design information, these test requirements can then be transformed into test cases, test oracles and test drivers. Test requirements help in devising the system test plan, in sizing the system test task and in planning appropriate resources early in the lifecycle.

Derivation of system test requirements in [10] is part of a system testing methodology called TOTEM. In TOTEM, the goal is to compare implementation against specification, hence the artifacts used are produced in the analysis stage. TOTEM uses the following UML diagrams as input artifacts:

- Use case diagram.
- Use case descriptions.
- Sequence or Collaboration diagrams.
- Class Diagram for application domain classes.
- A data dictionary that describes each class, method and attribute.
- Class invariants and operation contracts expressed in OCL.

TOTEM consists of eight steps (A1–A8) as illustrated in Fig. 2. A1 ensures testability. Testability is defined as the degree to which a model has sufficient information to support automatic test case generation. The next five steps are concerned with the derivation of test requirements. A2 to A5 derive test requirements from different artifacts. A6 merges all of them into one set and derives a test plan. A7 and A8 generate test cases and test oracles.

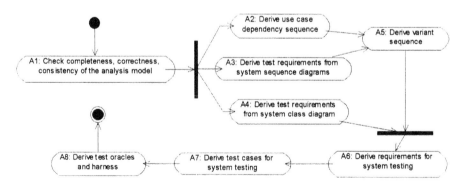

FIG. 2. TOTEM steps adapted from [10].

Briand et al. [10] address steps A2, A3 and A5. We will explain the approach by deriving test requirements for a part of our example course registration system. The following are a subset of registration system's functionalities that are used to demonstrate this technique:

- Add (Remove) students to (from) the system.
- Add (Remove) courses to (from) the system.
- Add (Remove) sections for each course.
- Register (Drop) courses for each student.

There is a sequential dependency among the use cases. For example, assuming that there are no students or courses when the system is first used, the Remove student (course) use case can only be executed after Add student (course) is executed at least once. This sequential dependency among use cases is used when test requirements are required. Sequential dependencies are represented with an activity diagram for each actor; vertices are use cases and edges are sequential dependencies. Figure 3 shows the activity diagram representing use-case sequences for actor registrar in the example.

Use-case parameters (both input and output) are listed in parentheses to show the dependencies between parameters during path execution. The activity diagram in Fig. 3 is converted to the directed graph of Fig. 4. The use-case sequence derivation is initiated by path derivation in the directed graph via a depth first search. Loops can cause infinite paths. To avoid this, loop iterations are limited to zero (if possible), one, an average value greater than one and a maximum value. Figure 5 shows the paths in the form of a tree derived from the directed graph in Fig. 4.

Next, dependencies among actual use-case parameters along the path need to be determined. For instance, going back to our example, in path AddCourse. AddSection.RemoveSection.RemoveCourse, the parameter cid for use case AddSection must be identical to parameter cid in AddCourse. These parameter dependencies are used to derive data flow information in use-case sequence executions, which is necessary for the generation of test data. These dependencies can be documented simply by representing the parameters in the sequence. These use-case sequences are called parameterized use-case sequences. Using the example path mentioned above, we obtain:

```
AddCourse(cid).AddSection(cid,secid)
.RemoveSection(cid, secid).RemoveCourse(cid)
```

Next, instantiated use-case sequences are derived by replacing parameters in each use-case sequence with a symbolic value. Several instances of one parameterized use-case sequence may be created depending on the number of objects participating in

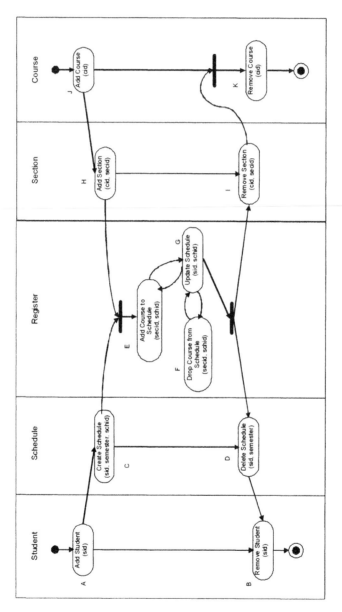

Fig. 3. Use-case sequential dependency diagram for the registrar.

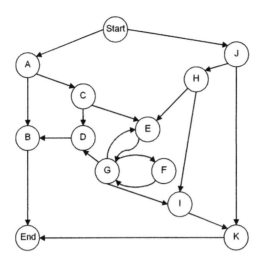

FIG. 4. Directed graph corresponding to Fig. 3.

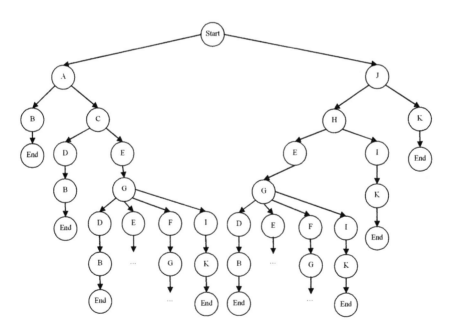

FIG. 5. Tree derived from the directed graph in Fig. 4.

the test. For instance, if the tester indicates one course and two sessions, the following instantiated use-case sequence is derived from the parameterized use-case sequence mentioned above:

```
S1:AddCourse(cid1).AddSection(cid1,secid1)
.RemoveSection(cid1,secid1).RemoveCourse(cid1)
S2:AddCourse(cid1).AddSection(cid1,secid2)
.RemoveSection(cid1,secid2).RemoveCourse(cid1)
```

All instantiated use-case sequences are combined into one set by finding common use cases across pairs of sequences. For each pair, all sub-sequences between each common pair of use cases are instantiated by generating a subset of all possible sequences resulting from combining the sub-sequences. For the example (S1, S2), a possible combined sequence is:

```
AddCourse(cid1).AddSection(cid1,secid1)
.AddSection(cid1,secid2).RemoveSection(cid1,secid2)
.RemoveSession(cid1,secid1).RemoveCourse(cid1)
```

Next, a sequence of use-case scenarios to be tested is derived. Use-case scenarios are represented by sequence diagrams. Thus, we need to have a sequence diagram describing the use-case scenarios for each use case in the test plan. Figure 6 represents the sequence diagram for the RemoveCourse use case. Each sequence diagram is represented as a regular expression in sum of products from where its alphabets are the public methods of the objects participating in the sequence diagram.

The '.' represents sequences of message calls, while '+' denotes alternative sequences. The following regular expression is the sum of product form for the sequence diagram in Fig. 6:

```
% TERM 1
diplayCourseList_RegistrarTerminal
.removeCourse_RegistrarTerminal
.displayError_RegistrarTerminal
.displayCourseList_RegistrarTerminal
+
% TERM 2
diplayCourseList_RegistrarTerminal
.removeCourse_RegistrarTerminal
.~course_Course.~section*_RegistrarTerminal
.updateSchedules_Schedule
.displayCourseList_RegistrarTerminal
+
```

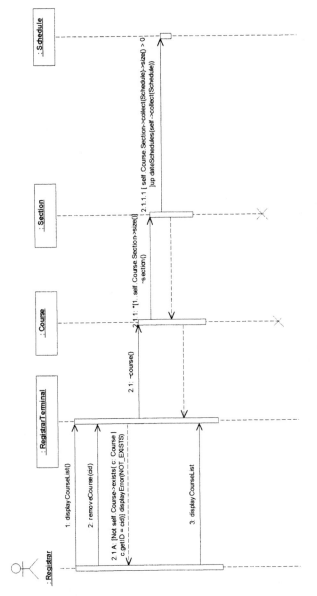

FIG. 6. Analysis sequence diagram for RemoveCourse.

% TERM 3
diplayCourseList$_{\text{RegistrarTerminal}}$
.removeCourse$_{\text{RegistrarTerminal}}$
.~course$_{\text{Course}}$.~section$^{*}_{\text{RegistrarTerminal}}$
.displayCourseList$_{\text{RegistrarTerminal}}$

Each term in the regular expression is associated with a number of conditions enabling or disabling its execution. These *path realization conditions* are expressed in OCL. The *path realization condition* for TERM 2 is:

[not self.Course->exists(c:Course-c.getID = cid)]

and

[self.Course.Section->collect(Schedule)->size()> 0]

After identifying the path realization condition for each term, the precise operation sequences need to be identified taking into account the actual number of iterations. The next step is to identify test oracles. Test oracles for each test sequence are derived from post conditions for each operation participating in the operation sequence. One way of deriving test oracles is to put assertions as pre/post conditions and class invariants at entry/exit points of each operation and raise an exception whenever they are violated.

Next, test requirements are formalized in the form of a decision table. The decision table consists of variants for each use case. Each variant corresponds to a path realization condition for one of the terms in the *sum of products* form of the regular expression of the sequence diagram. The decision table also has columns for initial conditions and the actions that are taken as a result of running the test cases. The actions correspond to system state changes and output messages being sent to actors. Steps A2, A3 and A5 are automated with a prototype tool.

3.1.2 Test Objectives and AI Planner-based Test Generation

Von Mayrhauser et al. [39] describe an approach to black-box test generation in which an artificial intelligence (AI) planner is used to generate test cases from test objectives derived from UML class diagrams. They developed a representation method at the application-domain level that allows for the statement of test objectives at that level and their mapping into a planner representation. A planner is an algorithm that finds a sequence of actions (i.e., plan) to reach a specified goal from the specified initial state, considering all the constraints and assumptions. Thus, AI planning is goal oriented and assuming that the test objectives are derived, the planner then generates a set of tests that achieves the test objectives.

The test objectives are described as sets of objects in terms of the states they can take on. Desired states for an object are determined through its attribute values and its links to other objects. The approach in [39] to derive test objectives is based on three steps:

1. Test objectives are derived for each class (and its instantiated objects) in the diagram separately, considering the class' relationship to other classes.
2. Test objectives are aggregated from those for individual classes (and their instantiated objects, subject to constraints through class relationships).
3. Test objectives are expressed as states of objects instantiated from classes depicted in the UML class diagram.

Test objectives are derived based on four parameters:

- number of instances of each class.
- properties for such instances.
- selections for how many instantiated objects should be brought into a particular state.
- states that they can take on.

Table III lists the building blocks for test objectives. Test objectives are derived by choosing one option for each of the four parameters. This leads to the 4-step process for deriving test objectives for a class:

1. Select a class, for which the states to be used as a test objective need to be identified.
2. Reduce the class diagram into a sub-diagram consisting of the classes and relationships that offer state information relative to the primary class. Table IV lists the detailed rules used to create sub-diagrams.
3. Select the set of instantiated objects for the primary class. This is a subset of the system configuration.
4. Determine the states for the set of instantiated objects. Select one or more of these states as the test objective.

TABLE III
BUILDING BLOCKS OF CLASS-BASED TEST OBJECTIVES

Number of Objects	Property	Participating Objects	States
1	p_1	All	s_1
$1 < n < max$	...	Sampling one	...
max	p_y	None	s_k

TABLE IV
RULES TO CREATE SUB-DIAGRAMS

1. Ignore aggregations. They are captured in persistent initial condition of the test generator.
2. Substitute subclasses of generalization relationships. Subclasses inherit attributes, methods and associations of the superclasses. There should be no superclasses or generalization relationships in the diagram following this step.
3. Keep only classes that are in (bi)directed associations with the primary class.
4. Reduce bi-directed associations with the primary class to directed associations with the primary class as the source class.
5. Remove any dangling associations (associations without classes on both sides).

Test objectives have two parts:

1. State information: desired goal states with regards to testing objects of this class.
2. Implications for necessary system configurations: how many objects could be instantiated.

Test directives guide the generation of test cases by converting the test objectives into a problem description. The AI Planner derives test sequences to achieve test objectives based on this problem description. The postprocessor converts them into executable test cases. A drawback of this approach is that when multi-object test objectives are aggregated, the space of possible test objectives can become intractably large.

Here, we will demonstrate the test objective derivation for the schedule class (as primary class) of the registration system. The classes and relationships that offer state-related information relative to class schedule are shown in the sub-diagram of Fig. 7. Note that according to the rules in Table IV, the sub-diagram in Fig. 7 should have included more classes (i.e., instructor, department and student), but since those classes do not offer any state-related information, there is no need to include them in the sub-diagram.

Figure 8 shows five test objectives derived specifically for the class schedule. The second column shows the number of instances of schedule participating in each test. For example in case of the first test objective, there are two instances of schedule (i.e., sch0000, and sch0001) participating in the test scenario. The third column shows the state-dependent attributes and their values. These attributes are used to gather information about the state of the object. The fourth column denotes the number of objects that should satisfy the test objective. Finally, the goal state of the primary class for each test objective is defined in the last column. For example, the following test objective is defined as the fourth test objective in Fig. 8: From the four schedules participating in the scenario, add a section of a course to two of the schedules where the section is full and there are already two other schedules in the reservation list of that section.

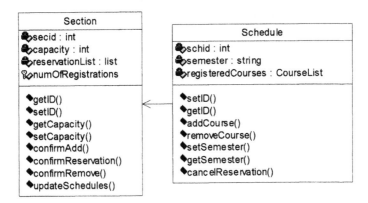

FIG. 7. Example of sub-diagram.

Test objective	Instantiation	Attribute	Participating Objects	State Value
1	Sch0000 Sch0001	registeredCourses.size > 0 Section.capacity = 0 Section.reservationList.size = 1	One	Empty schedule
2	Sch0010 Sch0011	registeredCourses.size = 0 self.Section.capacity = 0 self.Section.reservationList.size > 0	All	Pending course registration request
3	Sch0100 Sch0101	registeredCourses.size > 0 self.Section.capacity = 1 self.Section.reservationList.size = 0	One	
4	Sch0110 Sch0111 Sch1000 Sch1001	registeredCourses.size = 0 self.Section.capacity = 0 self.Section.reservationList.size = 2	Two	Schedule not empty
5	Sch1010	registeredCourses.size > 0 self.Section.capacity = 0 self.Section.reservationList.size = 3	All	Pending course removal request

FIG. 8. Example test objectives.

3.1.3 Test Generation from State Charts

Offut et al. [29] derive test cases from UML statecharts based on the type of the event. They defined four types of events: *call events, signal events, time events* and *change events*. They use change events as the basis for test generation. Test generation addresses four levels of test coverage: *(1) Transition coverage, (2) Full predicate coverage, (3) Transition pair coverage* and *(4) Complete sequence coverage.*

Each test case consists of the following: (1) an initial state, (2) a sequence of states to reach the initial state, (3) a sequence of testing steps and (4) a final state. A testing step is composed of the following components:

1. Action name: this is the name of the function(s) that is executed when the transition is traversed.
2. Clause name and value tuples: a value is assigned to each clause in the predicate on the transition.
3. Attribute name and value tuples: the before and after value of each attribute that is changed by the execution of the transition is denoted.
4. The next state: the state which follows the test step.

An automatic test generation tool (**UMLTest**) has been developed to support the process. A limitation of this work is that it does not support all the transitions in statecharts.

Figure 9 represents the statechart for the registration system. The four automatic tasks of the registration system are modelled using the five change events shown in this figure. Figure 10 shows the general test-set structure and example test sets for the registration system. Full predicate coverage has been used to derive these test sets. In

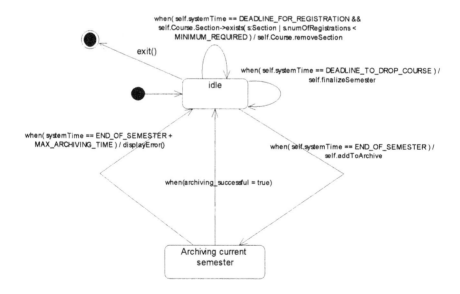

Fig. 9. Statechart diagram for registrarTerminal.

Standard Structure of a Test Set	Test Set = { *action_name*, *<clause_name, clause_value>**, [*before(attribute, value)*), *after(attribute, value)]**} Test Set = Test Case U {*init_state, prefix(s)*,($\sum$ *test set$_i$*) , *next_state*}
Full predicate coverage	T$_1$={idle, φ, self.removeSection, <self.systemTime, DEADLINE_FOR_REGISTRATION>, <self.Course.Section->exists(s:Section | s.numOfRegistrations, (MINIMUM_REQUIRED - 1)>, φ, idle} U {idle, φ, self.removeSection, <self.systemTime, DEADLINE_FOR_REGISTRATION>, <self.Course.Section->exists(s:Section | s.numOfRegistrations, MINIMUM_REQUIRED>, φ, idle} U {idle, φ, self.removeSection, <self.systemTime, (DEADLINE_FOR_REGISTRATION - 1)>, <self.Course.Section->exists(s:Section | s.numOfRegistrations, MINIMUM_REQUIRED>, φ, idle}
	T$_2$={idle, φ, self.addToArchive, <self.systemTime, END_OF_SEMESTER>, [before(status, "idle"), after(status, "archiving")], archiving current semester} U {idle, φ, self.addToArchive, <self.systemTime, END_OF_SEMESTER - 1>, [before(status, "idle"), after(status, "idle")], idle}

Fig. 10. Example of test case generated using [29].

the first example, the predicate consists of two clauses, hence we should have three different test cases. The initial state is the idle state. Since the system starts at the idle state, there are no prefix states (i.e., denoted by φ). The change event in this example executes when the deadline for registration has been reached and there exists at least one class that has less than the minimum number of students required for a course. As the transition is traversed, the removeSection function is executed. The next state of the system is again the idle state. This means that there is no change in the value of state attributes.

The second example demonstrates the transition of the system from the idle state as the initial state to the archiving state as the final state. In this case there is only one clause associated with the predicate. According to the full predicate coverage criterion this will result in two test cases. One that executes and one that fails to execute the transition. Note that in the first test case the before and after value of the `status` attribute is different as opposed to the second test case.

Briand et al. [9] propose a different approach for deriving test data from UML statecharts. When statecharts are tested, two consecutive steps need to be performed. First, a sequence of transitions (i.e., *transition test sequence*) needs to be defined. The second step is to assign values to arguments (i.e., parameter values) for events and actions in transitions. In the process of assigning values to test arguments, some constraints on test data need to be solved (e.g., the value of an event's parameter should be smaller than an attribute's value). Briand et al. focus on constructing a problem domain for the constraints involving the test data.

The input domain to this approach consists of a specific class to be tested, its statechart, its associated classes and a set of interacting statecharts belonging to some of the associated classes. The output is a set of constraints on the state of the system and on specific arguments for events and actions. For every path tested, the system state for each event/transition and also the input domain for parameters are automatically determined. Compared with the approach described in [29], this approach is less restrictive. This means that in addition to change events, other kinds of events are also dealt with. Furthermore, guards can involve attributes that are not necessarily of boolean type.

This methodology is based on normalization and analysis of event/action contracts and transition guards written with the object constraint language *(OCL)*. The concept of an invocation sequence tree *(IST)* is introduced: it defines the invocation chain caused by actions received by objects that have state-dependent behaviour. This tree-like data structure shows all possible invocation scenarios that may occur during the execution of a transition test sequence. Constraints are derived from the tree by normalizing *OCL* expressions. The normalization will support the analysis process in terms of constraint derivation and consistency checking among *OCL* expressions. An algorithm produces a sequence of sequences whose elements are constraints. This is done by traversing the *IST*, starting with the first trigger event. As a transition test sequence is traversed, all the associated constraints must be fulfilled.

We demonstrate this approach by using it to test the class `Section` of our registration system. The statechart of class `Section` is shown in Fig. 11. The set of classes that are associated with `Section` are: `Schedule`, `Instructor` and `Course`. Since `Instructor` and `Course` are stateless, the only interacting statechart is the `Schedule`'s statechart, which is represented in Fig. 12.

A transition test sequence has the following general form: `@state0@event0 [pred0]/actions0 @state1@event1[pred1]/actions1@...,` where `state0,state1,...` are the states of the statechart diagram; `event0, event1,...` are the input events; `pred0,pred1,...` are the predicates derived from the corresponding guard conditions and `actions0,`

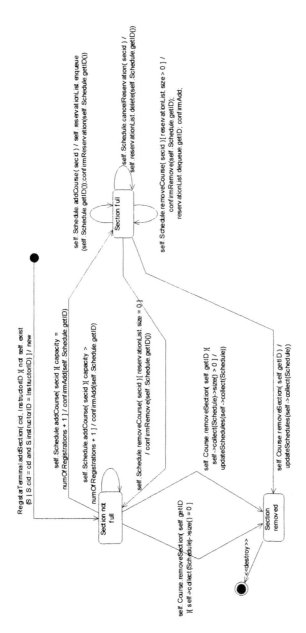

FIG. 11. Statechart diagram for class Section.

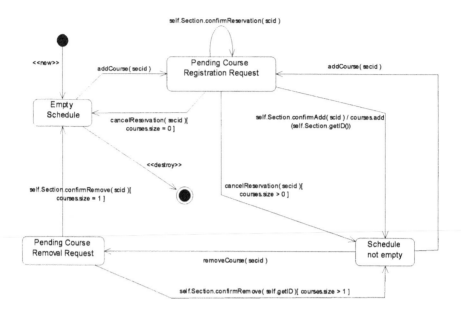

FIG. 12. Statechart diagram for class **Schedule**.

TABLE V
EXAMPLE OF A TRANSITION TEST SEQUENCE (TTS)

```
@section not full@self.Schedule.addCourse(s1)[true]
/self.confirmAdd(self.Schedule.getID)@section full
@self.Schedule.addCourse(s1)[true]
/self.reservationList.enqueue(self.Schedule.getID);
self.confirmReservation(self.Schedule.getID)
@section full@self.Schedule.cancelReservation(s1)
[self.reservationList.size > 0]/self.reservationList
.delete(self.Schedule.getID)@section full
@self.Schedule.removeCourse(s1)[true]/
self.confirmRemove(self.Schedule.getID);
@section not full@self.Course.removeSection(s1)
[true]/self.updateSchedules(
self->collect(Schedules));destroy;
```

actions1,... are the set of actions generated at each state upon receiving an
event if the guard condition is evaluated as being true. In case there is no predicate on
the transition, a **[true]** is written instead. Let us assume that we want to test the
transition test sequence (TTS) of Table V on an instance of **section**.

This sequence corresponds to the following scenario: The registrar adds a course (section **s 1**) to the schedule of a student. Then that section becomes full. The registrar tries to add the same section to another student's schedule. Since the section is full, the request is put into the reservation list. At this stage, a withdrawal of registration request (i.e., **cancelReservation**) is issued and in response the corresponding schedule is removed from the reservation list of the section. Registrar then removes section **s 1** from a student's schedule. Finally, the registrar decides to completely remove the section from the system. Although the section is not in full state, it is not empty, thus, the schedules of those students that are registered in the section are updated and then the section is removed. Figure 13 shows the IST for this scenario. The directed arc between $c5.1$ and $c5.2$ means

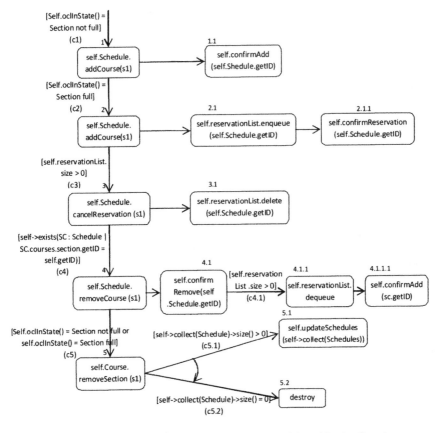

FIG. 13. Invocation Sequence Tree corresponding to Schedule and Section Statecharts.

that `self.updateSchedules(self->collect(Schedules))` and `destroy` are executed in sequence.

Each sub-scenario has at least one test constraint. The constraint is derived by propagating constraints that appear in the tree branches of the sub-scenario (i.e., invocation conditions for edges and post-conditions for nodes) onto the first edge of the sub-scenario (e.g., edge labelled (c5) for event (5) sub-scenario). The state of the system is changed by the execution of each operation in the IST. This means that for each operation, the pre- and post-conditions must be mapped to a condition that is meaningful in the pre-state of that operation. It is important to eliminate all local variables and query conditions in the OCL expressions before the constraint propagation process is started. This is done simply by replacing them with their actual values. The following is the test constraint derived for sub-scenario (5.2):

```
{[self->collect(Schedule)->size() > 0] or
[self->collect(Schedule)->size() > 0]} and
{[self.oclInState() = Section not full] or
[self.oclInState() = Section full]}
```

Latella et al. [24] proposed a formal testing framework for a behavioural subset of UML statechart diagrams *(UMLSDs)*. They also provide a way for effective automatic verification of testing equivalence of statecharts, based on existing techniques and tools. The approach converts statechart diagrams into Hierarchical Automata, which are then analysed. A drawback of this technique is that it does not consider history, action and activity states.

Latella et al. [16] extended this work and proposed a formal conformance testing relation and a non-deterministic test-case generation algorithm. The test-case generation algorithm generates test cases in a language that is a mix of process algebra (guarded action prefix, choice, and process definition/instantiation) and a simplified version of the lambda calculus. The main contribution of this work is to set the theoretical basis for test-case generation in a conformance testing setting. In order to use the proposed test-generation algorithm in practice, proper test-selection strategies are needed.

3.2   For UML

3.2.1   State Validation via Snapshot Creation

Gogolla et al. [17] propose an approach to validate system states. These states, called snapshots, are represented by object diagrams which consist of objects, attribute values for each object and links among objects. Snapshot validation is done using a tool called **USE,** which the authors developed earlier. The goal here is to facilitate the snapshot generation by defining the properties that they need to satisfy. For this purpose, Gogolla et al. developed A Snapshot Sequence Language (ASSL) that allows for the generation of desired snapshots by specifying their properties.

ASSL defines the sequence of operations that is needed to generate a snapshot with **USE**. In other words, the properties of snapshots are integrated with ASSL procedures. Test cases examine the system with respect to desired properties that need to be satisfied by class diagrams. Each property in the test case is specified using *dynamic* invariants as opposed to *present* invariants that need to be satisfied globally. Dynamic and present invariants are defined using OCL expressions. When a dynamic invariant is loaded into **USE**, it will make sure that there exists a snapshot that satisfies both the dynamic invariant and the present invariant. This is done by showing that there exists no valid snapshot that satisfies the present invariant as well as the negation of the dynamic invariant.

A test case in **USE** is a script that contains commands to generate the desired snapshot. Table VI shows the commands needed to generate a snapshot of the registration system that consists of one instance of every class in the registration system class

TABLE VI
EXAMPLE TEST CASE FOR **USE**

```
gen start regsys.assl generateCourse(1)
gen start regsys.assl generateSection(1)
gen start regsys.assl generateInstructor(1)
gen start regsys.assl generateDepartment(1)
gen start regsys.assl generateRegistrarTerminal(1)
gen start regsys.assl generateStudent(1)
gen start regsys.assl generateSchedule(1)
gen start regsys.assl generateCourseCatalog(1)
gen load student schedule.invs
gen load student department.invs
gen load student registrarterminal.invs
gen load schedule section.invs
gen load section inctructor.invs
gen load section course.invs
gen load course department.invs
gen load course coursecatalog.invs
gen load course registrarterminal.invs
gen load coursecatalog registrarterminal.invs
gen load instructor department.invs
gen load instructor registrarterminal.invs
gen load department registrarterminal.invs
gen start regsys.assl generateScheduleStudentLink(1)
```

diagram. For the sake of simplicity, we only considered the constraints on the association-end multiplicities as invariants. In other words, we have thirteen invariants i.e., one for each association link in the class diagram. The test case uses the ASSL procedures defined in `regsys.assl` to instantiate one instance of every class and then loads the invariants for each association link from the corresponding `.invs` file. Finally, the ASSL procedure `generateScheduleStudentLink(1)` asks **USE** to generate a snapshot that has one link between the schedule and student objects. **USE** automatically generates all the other links that are required based on the loaded invariants. Note that in our example if one of the objects e.g., Instructor, had not existed, no snapshot would have been generated. This is due to the fact that some of the invariants, namely, `section_inctructor.invs`, `instructor_department.invs` and `instructor_registrar-terminal.invs` require at least one instance of instructor to participate in the snapshot.

3.2.2 Test Execution via JAL

Trong et al. [38] introduced a testing approach in which executable forms of UML design models are exercised with test inputs generated from the class diagrams and activity diagrams. Later, the expected behaviour and the observed behaviour are compared and failures are reported. Their approach is supported by a prototype tool. Class diagrams are used to characterize a set of valid object configurations, while activity diagrams help define class operations. A Java-like Action Language *(JAL)* [28] is employed to describe the semantics of actions. The testing process begins with the introduction of the Design Under Test *(DUT)* into the testing system. *DUT* is transformed into Executable *DUT (EDUT)*. *EDUT* is a program that simulates the behaviour modelled in the *DUT*. *EDUT* contains two parts: a static structure representing the runtime configuration of the *DUT* and a simulation engine.

The static structure is derived from the class diagrams, while the simulation engine is generated from the activity diagrams. Test scaffolding is added to *EDUT* to perform failure checks *(TDUT)*. Test cases are implemented on the *TDUT* and results are reported by an observer class. Figure 14 illustrates an overview of this approach.

Figure 15 shows the activity diagram in the form of JAL specification for `addCourse` operation of the `RegistrarTerminal` class and the related partial class diagram. The corresponding *EDUT* is generated by combining information from the class diagram and activity diagram. In addition to instances of classes that are part of the test case, *EDUT* includes the following classes:

- `SetOfC`: Each instance of *SetOfC* maintains a collection of instances of class *C*. The purpose of this class is to take care of association-end multiplicities.

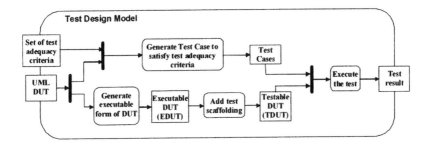

FIG. 14. Overview of the testing process.

- **TFactory**: This is a class that has public methods to create and destroy instances of every class and association in the class diagrams.

Figure 16 represents the *TDUT* for the **addCourse** operation in Fig. 15. The method **_addCourse** that is generated as part of *EDUT* is called from **addCourse**. Lines 2–8 and 10–15 are inserted to check pre- and post-conditions of the **addCourse** operation. The USE tool [17] is used to validate the objects' configuration and pre- and post- conditions of each operation. Any detected violation is reported as test failure.

Figure 17 shows a sample test case for operation **addCourse**. Test cases are represented by an abstract class called **TestDriver**. The **executeTest** is an abstract method in **TestDriver** which implements each test case. For each test case, **executeTest** is implemented as a method within class **TestDriverImpl,** which is a sub-class of **TestDriver. executeTest** has two parts: a prefix to create the start configuration and a sequence of system operation calls.

3.2.3 Testing Multiple Diagram Types

Most existing testing approaches for UML designs provide simple static analysis capabilities that can check model consistency. This can be done by validating structural views (e.g., class diagrams) against invariants represented by OCL expressions, or by validating behavioural views (e.g., sequence diagrams) against pre- or post-conditions represented by OCL expressions. However, these approaches do not validate dependencies between views. In other words, they do not address the problem of revealing inconsistencies among behavioural and structural views.

Pilskalns et al. [33, 34] address this problem by introducing a framework to test behavioural and structural aspects of UML designs by integrating the two views into a single representation called Testable Aggregate Model (TAM)[33] (its earlier version

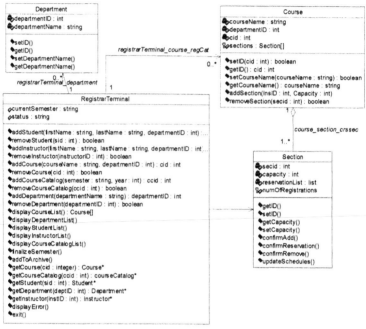

(a) Partial class diagram

```
[1]    Department dept;
[2]    dept = this.getDepartment(deptID);
[3]    if ( dept != null ) {
[4]        Course crs;
[2]        crs = _create_object_course();
[3]        crs.setCourseName(cname);
[4]        crs.setDepartmentID(deptID);
[5]        Section sec;
[6]          sec = _create_object_section();
[7]          _create_link_course_section_crssec(crs, sec);
[8]          return crs.getID();
[9]    }
[10]   else {
[11]       return -1;
[12]   }
```

(b) JAL specification for addCourse operation

FIG. 15. Partial DUT of Registrar System related to Add Course Scenario.

```
[1]   public int addCourse(string cname, int deptID){
[2]        try{
[3]             use.optEnter("openter   " + getUniqueName() + "  addCourse(" +
[4]             String.valueOf(cname) + "," + String.valueOf(deptID) + ")" );
[5]        }
[6]        catch(Exception e){
[7]             System.out.printIn(e.getMessage());
[8]        }
[9]        int _ret = _addCourse(cname, deptid);
[10]       try{
[11]            use.optExit( String.valueOf( _ret ));
[12]       }
[13]       catch(Exception e){
[14]            System.out.printIn(e.getMessage());
[15]       }
[16]       return _ret;
[17]  }  // EDUT code for addCourse() follows
[18]
[19]  private boolean _addCourse(string cname, int deptID){
[20]       Department dept;
[21]       dept = this.getDepartment(deptID);
[22]       if(dept != null){
[23]            crs = _factory()._create_object_course();
[24]            if(sec == null)
[25]                 reportError("Message sent to null object");
[26]            crs.setCourseName(cname);
[27]            crs.setDepartmentID(deptID);
[28]            sec = _factory()._create_object_section();
[29]            if(sec == null)
[30]                 reportError("Message sent to null object");
[31]            return crs.getid();
[31]       }
[32]       else{
[33]            return -1;
[34]       }
[35]  }
```

FIG. 16. Partial TDUT generated for the addCourse operation.

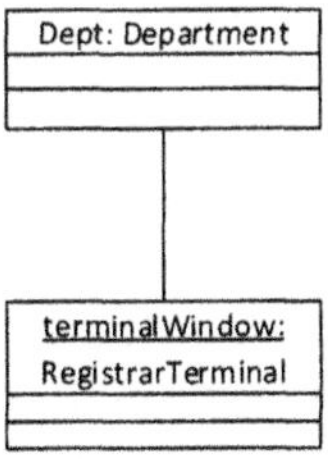

```
class TestDriverImpl extends TestDriver{
    boolean executeTest(){
        // Prefix
        RegistrarTerminal terminalWindow = factory.createRegistrarTerminal();
        Department Dept = factory.createDepartment();
        Factory.create_link_registrarTerminal_department(terminalWindow, Dept);
        Dept.setID(1);
        //Sequence of system operation calls
        terminalWindow.addCourse("SE", 1);
    }
}
```

Dept: Department

terminalWindow:
RegistrarTerminal

(a) Start Configuration

(b) TestDriverImpl::executeTest():
A test case implementation

FIG. 17. A sample test case.

was called Object Method Directed Acyclic Graph (*OMDAG*))[34]. They provide a framework to generate and execute test cases using TAM and to validate test results by comparing them against OCL expressions.

The TAM is constructed by combining the behavioural information of sequence diagrams with the structural information of class diagrams. This aggregation of sequence and class diagrams makes this approach different from [17] as it allows validation of multiple types of diagrams at the same time, allowing for the effective testing for cross-diagram defects. The approach consists of the following steps:

1. Build TAM using UML models.

    (a) Construct a Directed Graph (DG) from each sequence diagram.
    (b) Construct Class and Constraint Tuples (CCT) from class diagram and OCL expressions.
    (c) Combine DG and CCT into TAM.

2. Determine input model and generate test cases.

    (a) Determine which attributes need partitioning.
    (b) Partition attributes with domain analysis [7] to generate test cases.

3. Execute the tests.

    (a) for each test:

        i. record potential faults.
        ii. validate test results.

Going back to our registration system example, Fig. 18 shows the sequence diagram for a scenario where a course is added and then a course (which might be different) is removed from the system. We will use this sequence diagram as an input to demonstrate this methodology. At the very first step, a DG needs to be derived from the sequence diagram. The construction of DG starts by traversing the first message in the sequence diagram and creating its corresponding vertex. In general, if m_i and m_j are two messages in the sequence diagram and v_i and v_j be the corresponding vertices, an edge is added from v_i to v_j if it is possible to execute m_j directly after m_i.

A DG is represented by the tuple $G = \langle V, E, s \rangle$, where V is a set of vertices, E is the set of edges and s is the starting vertex. A vertex in DG can be a simple vertex representing a message or a sub-DG representing a *combined fragment*, hence representing several levels of abstraction. *Combined fragments* allow the developer to describe the control flow of messages with conditions. In the context of [33], three kinds of combined fragments are considered. These are *option* (i.e., 'if' statement), *alternative*

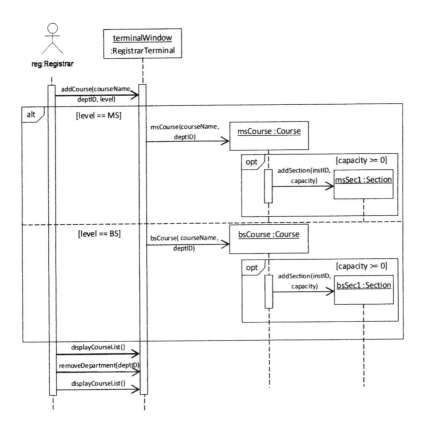

Fig. 18. Example of sequence diagram.

(i.e., 'switch' statement) and *loop*. The loop fragment may contain a boolean guard condition, as well as a minimum and maximum number of iterations.

The DG corresponding to the sequence diagram in Fig. 18 is represented in Fig. 19. This graph has four Sub-DGs which correspond to the two *alternative* constructs. Each sub-DG, $v_{sub} = \langle$ [boolean, min, max], $\langle V, E, s \rangle \rangle$, is composed of a DG and a guarded boolean expression. The boolean expression indicates the condition that must be satisfied for sub-DG to be traversed and the min and max values indicate how many times it should be traversed. Although the return messages of each function call are not shown in Fig. 18, there is a vertex present for each of them in the corresponding DG.

In general, each message vertex, v, is defined by the tuple $v = \langle o, m, \textit{lifeline}, \textit{ARGS}, c \rangle$, where o is an object calling m, m is the message, *lifeline* classifies an object as *new* if it is being created, *deleted* if it is being deleted and *exists* otherwise.

Directed Graph

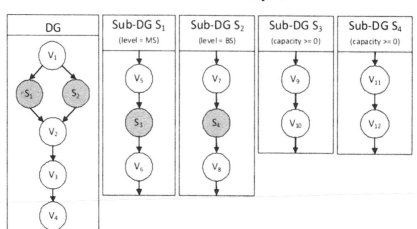

V$_1$<reg, addCourse(), exists, {<string, courseName, null>, <int, deptID, null>, <int, level, null>}, Registrar>
S$_1$<[level==MS]
    V$_5$<terminalWindow, msCourse, new, {<string, courseName, null>, <int, deptID, null>},
      RegistrarTerminal>
    S$_3$<[capacity >= 0]
        V$_9$<msCourse, addSection(), new, {<int, instID, null>, <int, capacity, null>}, Course>
        V$_{10}$< msSec1, return(), exists, {null}, Section >
    V$_6$ <msCourse, return(), exists, {null}, Course> >
S$_2$<[level==BS]
    V$_7$<terminalWindow, bsCourse, new, {<string, courseName, null>, <int, deptID, null>},
      RegistrarTerminal>
    S$_4$<[capacity >= 0]
        V$_{11}$<bsCourse, addSection(), new, {<int, instID, null>, <int, capacity, null>}, Course>
        V$_{12}$< bsSec1, return(), exists, {null}, Section >
    V$_8$ <bsCourse, return(), exists, {null}, Course> >
V$_2$<reg, displayCourseList(), exists, {null}, Registrar>
V$_3$<reg, removeDepartment(), exists, {<int, deptID, null>}, Registrar>
V$_4$<reg, displayCourseList(), exists, {null}, Registrar>

FIG. 19. Directed graph corresponding to sequence diagram of Fig. 18.

ARGS is a set of argument tuples and *c* the class name of the instance *o*. The *ARGS* tuple is composed of ⟨*type*, *name*, *value*⟩, where the *type* is the argument type, the *name* is the argument name and the *value* is any assigned value. For example, the first vertex in Fig. 19 represents the message a d d C o u r s e in Fig. 18. The calling object of addCourse is reg. The object registrarTerminal already exists. The arguments for addCourse are courseName of type string, deptID of type int and level of type int. Finally, the class name of the calling object reg is Registrar.

Constraint Class Tuples (CCTs) contain structural and constraint information. Class diagrams and OCL expressions are used to derive CCTs. OCL expressions contain pre/post conditions as well as invariants. OCL invariants can represent association and multiplicity information among classes. Constraint Class Tuples consist of a class name, attributes from class and superclasses (if applicable), operations for the class and superclasses and OCL information for both the attributes and the operations (e.g., pre/post conditions). A CCT of a class c has the form:

$$CCT(c) = \langle \{\langle ParentCCT \rangle\}, \{\langle Attribute \rangle\}, \{\langle Operation \rangle\}, \{[invariant]\} \rangle$$

where c is the class name, $\{\langle ParentCCT \rangle\}$ is a set consisting of parent class CCTs for each parent class of c, $\{\langle Attribute \rangle\}$ is a set of attribute tuples with constraints, $\{\langle Operation \rangle\}$ is a set of operation tuples with constraints and *invariant* is a set of constraints at the class level (e.g., number of instances). The *Attribute* tuple is defined as follows:

$$Attribute = \langle attributename, attributetype, visibility, invariant, \langle CCT \rangle \rangle.$$

The *Operation* tuple is defined as follows:

$$Operation = \langle name, returntype, visibility, pre\_condition, post\_condition,$$
$$\langle Parameters \rangle \rangle.$$

Figure 20 shows an annotated CCT for the class Course. The final step in building the aggregate model is to combine CCTs and DGs. This is done by replacing class name c in the DGs with their corresponding CCTs. While CCTs and DGs are being combined, static evaluation tests the consistency of methods and parameters.

Test cases consist of values for variables or attributes that enable traversing a path in the TAM. Thus, variables and attributes that are present in conditional statements constitute the input model. The set of values that they can obtain defines the input domain. The following steps define the input model:

1. Identify the set of variables that occur in conditions.
2. Determine the range based on type of variable. Use one of the combinatorial techniques in [7] to determine partitions and combinations of partitions.
3. Select test values based on the combinatorial techniques used in step 2.

Table VII shows the input model corresponding to the sequence diagram of Fig. 18. The size of the complete test set is $3 \times 6 = 18$, but since many of the test cases are

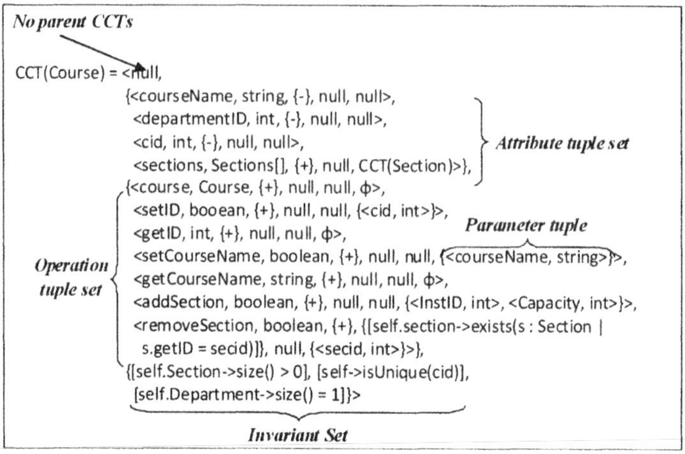

No parent CCTs

CCT(Course) = <null,
        {<courseName, string, {-}, null, null>,
         <departmentID, int, {-}, null, null>,
         <cid, int, {-}, null, null>,
         <sections, Sections[], {+}, null, CCT(Section)>},   } Attribute tuple set
        {<course, Course, {+}, null, null, φ>,
         <setID, booean, {+}, null, null, {<cid, int>}>,
         <getID, int, {+}, null, null, φ>,   Parameter tuple
Operation   <setCourseName, boolean, {+}, null, null, {<courseName, string>}>,
tuple set   <getCourseName, string, {+}, null, null, φ>,
         <addSection, boolean, {+}, null, null, {<InstID, int>, <Capacity, int>}>,
         <removeSection, boolean, {+}, {[self.section->exists(s : Section |
         s.getID = secid)]}, null, {<secid, int>}>,
        {[self.Section->size() > 0], [self->isUnique(cid)],
         [self.Department->size() = 1]}>

                    Invariant Set

FIG. 20.   Annotated CCT(course).

replicas of each other, we omitted those test cases. In general, in addition to ON and OFF boundary values, a typical value needs to be included. Test execution traverses the TAM using the generated test cases derived from the input model. To validate the results, the changes to the system during test execution must be recorded. This is facilitated by the use of an instance and trace table. The instance table keeps count of the number of instances for each class. The trace table records each message, each object and every attribute assigned values. The instance table is updated as execution proceeds for the class type of the object as well as all of its super classes (if available). For each conditional vertex v_i attribute values are assigned and recorded based on the executed test case. Tables VIII and IX show the trace tables for test cases T1 and T2, respectively.

With the trace table and instance table, two types of faults can be revealed. The first type is an *OCL fault*. It occurs when states recorded in the execution trace or instance tables violate OCL constraints. Table VIII shows an OCL fault. This is due to the OCL invariant that requires each section to have a capacity of at least one instead of zero. The second type is classified as *path fault*, where a path may not be traversable or may not exist. This can be caused by calling a private, abstract or non-existing operation. An example of this kind of error is demonstrated in Table IX. This error is caused because there is no path sequence to handle PHD course creation.

In addition to detecting OCL and path faults, the instance table can be used to detect association end multiplicity violations. For example, Table X reveals this kind of error. On the basis of the class diagram of the registration system, an instance of Course should always be associated with one department. After the execution of

TABLE VII
INPUT MODEL

Conditions	Boundary	T1	T2	T3	T4	T5	T6	T7	T8	T9
				Test Cases						
(level == MS)	ON	MS			MS		MS			
	OFF		PHD						PHD	
(level == BS)	ON			BS		BS		BS		
	OFF									PHD
(capacity ≥ 0)	ON	0				0			0	
	OFF		−1		−1			−1		
	TYPICAL			30			30			30

TABLE VIII
TRACE TABLE FOR T1

Calling Object	Operation Call	Attribute Values
Reg	addCourse	courseName = "SE", dep-tID = 1, level = MS
terminalWindow	msCourse	courseName = "SE", dep-tID = 1
MsCourse	addSection	instID = 1, capacity = 0 (OCLfault)

TABLE IX
TRACE TABLE FOR T2

Calling Object	OperationCall	Attribute Values
Reg	addCourse	courseName = "SE", dep-tID = 1, level = PHD (Path fault)

TABLE X
INSTANCE TABLE AFTER
EXECUTION OF T6

Class	Number
RegistrarTerminal	1
Course	1
Section	1
Department	0

T6, this constraint is violated. A suggested solution is to delete all courses associated with a department before removing the department. Pilskalns et al. also provide a prototype tool called **AdaptUML**[23], which automates the test generation, execution and validation.

4. Conclusions

This chapter explored the existing approaches to test implementations against their design, as well as testing approaches to test the design itself. Not surprisingly, most work centered around the most commonly used UML design artifacts, i.e., Class Diagrams, Sequence Diagrams and State Charts. Some approaches either require or prefer the use of OCL to diagnose inconsistencies, provide partial test oracles and locate faults. Testing criteria have been defined for Class Diagrams, Sequence/Collaboration Diagrams, and State Charts. We also provided a set of properties that reasonable UML design testing criteria should exhibit. Many of the techniques on which we reported are not (fully) tool supported. Some require information that is not included in UML artifacts (e.g., partitioning of the input model for test generation purposes). Most of the existing techniques provide testing methods for single types of UML diagrams. Only one technique allows for cross-diagram analysis by combining structural and behavioural diagrams into one notation [33]. None of the testing techniques goes beyond functional testing to evaluate other important design properties like performance. Empirical and experimental validation of the techniques is extremely limited.

On the basis of the current limitations of UML design testing, we recommend the following:

- Include testing techniques that span a larger number of UML diagrams types, including interfaces, components and phases (from business use case and analysis models to deployment and implementation models).
- Provide more tools and increase the degree of test automation.
- Provide approaches that allow cross-diagram analysis by combining structural and behavioural models to allow testing of more subtle interactions.
- Integrate multiple techniques for a more powerful approach. For example, Pilskalns et al. [33] technique could be strengthened by combining it with the technique of Trong et al. [38]. This would allow for actual test execution via the JAL, instead of the more limited symbolic execution. It would also need less OCL. Trong et al. [38] already make use of Gogolla's tool [17].
- Extend testing to include assessment of performance, security, fault tolerance, etc. Currently, a large body of work has been carried out to assess UML design performance [3, 4, 6, 11–13, 25, 27, 32, 37, 42]. Similarly, there are papers on assessing various security properties [5, 14, 20–22, 26, 31]. However, from an evaluator's point of view, it would be preferable if one did not have to work with different representations for each type of analysis. Again, an integrated representation that can be analysed with different methods would make this task easier.

Grassi et al. [18] suggest transforming a series of different design notations into an intermediate model that includes a model extractor for each target analysis (e.g., Petri net Analysis, Markov Processes, etc.). While this approach addresses performance and reliability analysis rather than testing, it follows the same philosophy as that of Pilskalns et al. [33], i.e., integrating different models for a more powerful analysis.

- Provide more empirical and experimental evaluation. There are over 25 years of testing technique experiments for code [19] including laboratory studies, formal analysis, controlled experiments and field studies. Given that, existing techniques used to test with UML designs are comparatively recent, we still have very little evidence with regards to the nature of common faults in designs, their frequency and the ability of the design test techniques to find them.

REFERENCES

[1] Abdurazik A., and Offutt A. J., 2000. Using UML collaboration diagrams for static checking and test generation. In *UML 2000: Proceedings of the 3'rd International Conference on UML*, pp. 383–395.

[2] Andrews A., France R., Ghosh S., and Craig G., October 2000. Test adequacy criteria for UML design models. *Journal of Software Testing, Verification, and Reliability*, **13**(2):95–127.

[3] Balsamo S., Di Marco A., Inverardi P., and Simeoni M., May 2004. Model-based performance prediction in software development: a survey. *IEEE Transactions on Software Engineering*, **30**(5):295–309.

[4] Balsamo S., and Marzolla M., Performance evaluation of uml software architectures with multiclass queueing network models. In *WOSP '05*.

[5] Basin D., Doser J., and Lodderstedt T., 2006. Model driven security: from UML models to access control infrastructures. *ACM Transactions on Software Engineering and Methodology*, **15**(1):39–91.

[6] Bernardi S., Donatelli S., and Meseguer J., 2002. From uml sequence diagrams and statecharts to analyzable petri nets. *Proceedings of the 3rd International Workshop on Software and Performance (WOSP)*.

[7] Binder R., 1999. *Testing Object-Oriented systems models*. Addison-Wesley.

[8] Booch G., Rumbaugh J., and Jacobson I., 2005. *The Unified Modeling Language User Guide – 2nd Edition*. Addison-Wesley.

[9] Briand L. C., Cui J., and Labiche Y., 2003. Towards automated support for deriving test data from UML statecharts. In *UML '03: Proceedings of the 6'th International Conference on UML*, pp. 249–264.

[10] Briand L. C., and Labiche Y., 2001. A UML-based approach to system testing. In *UML '01: Proceedings of the 4'th International Conference on UML*, pp. 194–208, Springer-Verlag, London, UK.

[11] Canevet C., Gilmore S., Hillston J., Prowse M., and Stevens P., 2002. Performance Modelling with Uml and Stochastic Process Algebra.

[12] Cortellessa V., and Mirandola R., Deriving a queueing network based performance model from uml diagrams. *Proceedings of the 2nd IEEE Workshop on Software and Performance (WOSP 2000)*.

[13] D'Ambrogio A., 2005. A model transformation framework for the automated building of performance models from uml models. *Proceedings of the 5th International Workshop on Software and Performance (WOSP)*.

[14] Georg G., Ray I., and France R., 2002. Using aspects to design a secure system. In *ICECCS '02: Proceedings of the 8'th IEEE International Conference on Engineering of Complex Computer Systems*, pp. 117–126, IEEE Computer Society, Washington, DC, USA.

[15] Ghosh S., France R., Braganza C., Kawane N., Andrews A., and Pilskalns O., 2003. Test adequacy assessment for UML design model testing. In *ISSRE '03: Proceedings of the 14'th International Symposium on Software Reliability Engineering*, pp. 332–343, IEEE Computer Society, Washington, DC, USA.

[16] Gnesi S., Latella D., and Massink M., 2004. Formal test-case generation for UML statecharts. In *ICECCS '04: Proceedings of the 9'th IEEE International Conference on Engineering Complex Computer Systems Navigating Complexity in the e-Engineering Age (ICECCS '04)*, pp. 75–84, IEEE Computer Society, Washington, DC, USA.

[17] Gogolla M., Bohling J., and Richters M., 2003. Validation of UML and OCL models by automatic snapshot generation. In *UML '03: Proceedings of the 6'th International Conference on UML*, pp. 265–279.

[18] Grassi V., Mirandola R., and Sabetta A., 2005. From design to analysis models: a kernel language for performance and reliability analysis of component-based systems. In *WOSP '05: Proceedings of the 5'th international workshop on Software and performance*, pp. 25–36, ACM Press, New York, NY, USA.

[19] Juristo N., Moreno A., and Vegas S., March 2004. Reviewing 25 years of testing technique experiments. *Empirical Software Engineering Journal*, 9(1/2).

[20] Jürjens J., 2002. UMLsec: extending UML for secure systems development. In *UML '02: Proceedings of the 5'th International Conference on the Unified Modeling Language*, pp. 1–9, Springer, Berlin, Germany.

[21] Jürjens J., 2004. *Secure Systems Development with UML*, Springer-Verlag, London, UK.

[22] Jürjens J., 2005. Sound methods and effective tools for model-based security engineering with UML. In *ICSE '05: Proceedings of the 27th International Conference on Software Engineering*, pp. 322–331.

[23] Knight A. S., August 2005. ADAPTUML: A Tool for Evaluating UML Designs. Master's thesis, Washington State University, Pullman.

[24] Latella D., and Massink M., 2001. A formal testing framework for UML statechart diagram behaviours: From theory to automatic verification. In *HASE '01: The 6'th IEEE International Symposium on High-Assurance Systems Engineering*, pp. 11–22, IEEE Computer Society, Washington, DC, USA.

[25] Lindemann C., Thummler A., Klemm A., Lohmann M., and Waldhorst O. P., 2002. Performance analysis of time-enhanced uml diagrams based on stochastic processes. *WOSP '02*.

[26] Lodderstedt T., Basin D., and Doser J., 2002. SecureUML: a UML-based modeling language for model-driven security. In *UML '02: Proceedings of the 5'th International Conference on the Unified Modeling Language*, pp. 426–441, Springer, Berlin, Germany.

[27] Lopez-Grao J. P., Merseguer J., and Campos J., From uml activity diagrams to stochastic petri nets: application to software performance engineering. In *Workshop on Software and Performance*.

[28] Mellor S. J., and Balcer M. J., 2002. *Executable UML: A Foundation for Model Driven Architecture*. Addison-Wesley Professional.

[29] Offutt A. J., and Abdurazik A., 1999. Generating tests from UML specifications. In *UML '99: Proceedings of the 2'nd International Conference on UML*, pp. 416–429.

[30] Parrish A., and Zweben S. H., 1991. Analysis and refinement of software test data adequacy properties. *IEEE Transactions on Software Engineering*, 17(6):565–581.

[31] Petriu D. C., Woodside C. M., Petriu D. B., Xu J., Israr T., Georg G., France R., Bieman J. M., Houmb S. H., and Jürjen J., 2007. Performance analysis of security aspects in uml models. In *WOSP '07: Proceedings of the 6'th International Workshop on Software and Performance*, pp. 91–102, ACM Press, New York, NY, USA.

[32] Petriu D. C., and Wang X., 2000. From uml description of high-level software architecture to lqn performance model. In M. Muench M. Nagl, A. eds. *Computer Performance Evaluation – Modelling Techniques and Tools*, number 2324, pp. 159–177.

[33] Pilskalns O., Andrews A., Knight A., Ghosh S., and France R., August 2007. UML design testing. *Journal of Information Science and Technology*, **19**(8):192–212.

[34] Pilskalns O., Andrews A. A., Ghosh S., and France R. B., 2003. Rigorous testing by merging structural and behavioral UML representations. In *UML '03: Proceedings of the 6'th International Conference on UML*, pp. 234–248.

[35] Rumbaugh J., Jacobson I., and Booch G., 2005. *The Unified Modeling Language Reference Manual*. Addison-Wesley.

[36] Scheetz M., von Mayrhauser A., France R., Dahlman E., and Howe A. E., 1999. Generating test cases from an OO model with an AI planning system. In *ISSRE '99: Proceedings of the 10'th International Symposium on Software Reliability Engineering*, pp. 250–259, IEEE Computer Society, Washington, DC, USA.

[37] Smith C. U., and Williams L. G., Performance evaluation of software architectures. In *Proc. 1st International Workshop on Software and Performance (WOSP 1998)*.

[38] Trong T. D., Kawane N., Ghosh S., France R., and Andrews A., 2005. A tool-supported approach to testing UML design models. In *ICECCS '05: Proceedings of the 10'th IEEE International Conference on Engineering of Complex Computer Systems*, pp. 519–528.

[39] von Mayrhauser A., France R., Scheetz M., and Dahlman E., 2000. Generating test-cases from an object-oriented model with an artificial-intelligence planning system. *IEEE Transactions on Reliability*, **49**(1):26–36.

[40] Weyuker E., 1988. The evaluation of program-based software test data adequacy criteria. *Communications of the ACM*, **31**(6):668–675.

[41] Weyuker E. J., 1986. Axiomatizing software test data adequacy. *IEEE Transactions on Software Engineering*, **12**(12):1128–1138.

[42] Woodside C. M., Petriu D. C., Petriu D. B., Shen H., Israr T., and Merseguer J., Performance by unified modified analysis (puma). In *Proceedings of Workshop on Software and Performance*.

Balancing Transparency, Efficiency and Security in Pervasive Systems

MARK WENSTROM, ELOISA BENTIVEGNA*
AND ALI R. HURSON

*Department of Computer Science and Engineering
Pennsylvania State University
University Park, PA 16802
wenstrom@cse.psu.edu, hurson@mst.edu*

**Department of Physics*
*Pennsylvania State University
University Park, PA 16802
bentiveg@phys.psu.edu*

Abstract

This chapter surveys pervasive computing with a focus on how its constraint for transparency affects issues of resource management and security. The goal of pervasive computing is to render computing transparent, such that computing resources are ubiquitously offered to the user and services are proactively performed for a user without his or her intervention. The task of integrating computing infrastructure with everyday life without making it excessively invasive brings about trade-offs between flexibility and robustness, efficiency and effectiveness, as well as autonomy and reliability. While efficiency in resource management is not the primary goal of pervasive computing, it should be considered in order to best utilize a limited set of resources (bandwidth, computing, etc.) so as to avoid congestion and creation of a visible and distracting bottleneck in the eyes of the user. As solutions to efficiently manage the resources in a pervasive computing environment, three techniques will be examined: the distributed caching and sharing of data between mobile hosts, the broadcasting of services by public service providers and the ability for mobile hosts to adaptively adjust the quality of offered services. Likewise, security is often an afterthought in many computing projects, though it should be of high consideration in a pervasive environment

ADVANCES IN COMPUTERS, VOL. 73
ISSN: 0065-2458/DOI: 10.1016/S0065-2458(08)00403-8

99

where users share public resources to operate on private data. Specifically, how can a user be authenticated in this environment with minimal or no user intervention? Solutions such as single sign-on via smartcards and biometrics will be examined to carry out authentication in a pervasive environment. As the feasibility of ubiquitous computing and its real potential for mass applications are still a matter of controversy, this chapter will look into the underlying issues of resource management and authentication to discover how these can be handled in a least invasive fashion. The discussion will conclude with an overview of the solutions proposed by current pervasive computing efforts, both in the area of generic platforms and for dedicated applications such as pervasive education and healthcare.

List of Abbreviations

FMR	false match rate
FNM	false non-match rate
FPS	frames per second
FR	read frequency
LRU	least recently used

MTBR	mean time between reads
MTBU	mean time between updates
P2P	peer-to-peer
PDA	personal digital assistant
PDF	probability density function
PM	probability that an object has been modified
PNM	probability that an object has not been modified
QoS	quality of service
SDS	secure discovery service
STDV	standard deviation
TGS	ticket granting service
TSP	travelling salesman problem

1. Introduction

Pervasive computing explores the task of integrating technology with an environment, such that a multitude of computing devices are available to proactively perform services for each user, thereby lightening the user's workload. It has been pointed out [1] that pervasive systems constitute the third wave in computing, after the main frame era (one computer, many users) and the personal computer era (one computer, one user). Pervasive computing is the next natural step in order to set a single user in control of several computing elements. It should be noted that occasionally, the literature has used the term 'ubiquitous computing' and 'pervasive computing' interchangeably. In this article, however, we are making a distinction between the two: Pervasive refers to the invisibility and proactivity where the computer dissolves into the fabric of the surroundings, while ubiquity refers to the availability. In other words, ubiquitous computing facilitates a better pervasive computing.

In a pervasive environment, a user should always have access to computing resources, whether the user is mobile or stationary; thus it is assumed that each user is in control of a personal mobile device (e.g., a PDA or laptop). One's device can be used to connect to a multitude of resources: it can wirelessly tap into an access point for connectivity to the global Internet; similarly, the device can tap into a local access point for connectivity to a local network (e.g., a university or office network); on the other hand, the device may not be able to tap into a structured network, but may be able to join an ad hoc network of wireless devices in order to utilize those resources.

As of the year 2007, the aforementioned forms of connectivity and resource sharing are widely available. The field of mobile computing has provided some of the framework for pervasive computing, as mobile devices currently give users access to computing resource at all times. Not only does pervasive computing require constant access to resources, but it also requires that technology be seamlessly and invisibly integrated into the lives of its users. Thus, the field of artificial intelligence must also be employed, such that one's mobile device can predict the desires of its user and can independently carry out services for this user. This implies that a mobile device must be aware of its surrounding context and must be able to locate and call upon remote resources to carry out its user's intent. Consequently, the field of distributed computing must be employed to provide techniques to divide out computations to remote resources; this is especially important to a pervasive environment, as the environment consists of a variety of computing elements, ranging from powerful servers to resource-constrained mobile devices.

These areas of computing can be adapted, fused and ameliorated to achieve the goal of pervasive computing: balancing proactivity of services and transparency of operation in order to saturate an environment with computing agents that automate the trivial daily tasks of life (e.g., transferring one's lecture notes from a PDA to a workstation), leaving humans free to focus on the high-level tasks (e.g., delivering a lecture). In other words, the focus of one's action is intended to be the high-level task rather than the technology.

Notwithstanding the availability of the required technology, true pervasive computing environments have not been realized, as only prototypes and theoretical designs have been developed by the research community. A major open field is related to the fact that pervasive computing faces the delicate issue of which choices can be delegated to the system (in the form of local clients, neighbouring peers or a central server) and which must be imperatively performed by the user. The goal of pervasive computing is obviously to maximize the former and minimize the latter. This often demands for smarter algorithms, architectures and technologies than those that are presently available. In order to create a system that proactively issues tasks, yet remains mostly transparent from the user, two broad-range issues, with ramifications and follow-ons need to be addressed:

- **The computing agents need to be able to predict the user's intent based on history and context awareness**. Predicting a user's intent implies that the services performed for a user should be desired by the user. Thus, the system must intelligently decide what services a user desires at each moment. For example, a user commuting to work will want to know if there is traffic along his route, and in this event he may need an alternate route. In the case where there is no traffic, the user should not be notified. Therefore, the system should know to check for traffic along the user's normal route, and calculate alternate routes in the event of traffic and relay this information to the user, and in the case where there is no traffic the user should not be bothered.

- **A reliable way of integrating all the computing agents into a seamless entity needs to be designed.** The requirement for a transparent system brings about difficult questions and issues that need to be addressed. First, what is an appropriate mobile device for a user? Should this device be a full-size laptop, capable of running applications and providing standard I/O? Or should the device be a thin client, which only provides I/O and outsources heavy computation to resource-abundant machines? As more features are supported by a mobile device, more resources are needed on the device, which implies a physically larger device. Yet, a physically large device violates the transparency requirement. Ideally, these devices would be wearable computers, implying that such a device can be carried as if it was a personal accessory, such as a wallet or wristwatch. Since a small device lacks computational power and battery life, the device must outsource its workload to more capable machines in order to provide the same features of larger devices. Furthermore, the issue of security of operation, when combined with the transparency requirement, raises additional caveats. First, the system needs to be endowed with user authentication procedures that grant a sufficient degree of security while at the same time retaining the invisibility feature of a pervasive application. Second, only authorized agents should have the capability of initiating processes for the user; a framework to scan and filter the available services will then have to be integrated into the system.

In this chapter, we will focus on two issues that are crucial for the design of a proactive yet transparent system, in which mobile devices outsource tasks and services to remote machines. First, we will describe techniques for resource management in a pervasive environment, considering a realistic assumption that the proactive issuing of tasks may overwhelm the available resources and may bring about distraction to the user. Second, we will present the issue of authentication and recognition of users and services, again focusing on solutions that minimize the amount of human–machine interaction but still provide the required level of security. We conclude our discussion with the review of a spectrum of current pervasive computing projects.

2. Resource Management

As outlined in the Introduction, a reasonable strategy to reduce the computational burden of small devices in a pervasive environment is to allow those clients to outsource their workload to larger machines. This outsourcing of work brings about its own issues in the realm of transparency.

The machines that take on the outsourced work are known as surrogates. A surrogate may be a standard workstation, a server-grade machine or even a cluster of servers. A mobile device may outsource a computationally intensive task to a surrogate through some type of remote procedure call. Thus, the mobile device could act as a thin client, whereby the surrogate runs a user's applications and the mobile device is only used for I/O. For simplicity, this article will group all types of outsourced tasks under the label of *services*. Therefore, the servers and surrogates processing the clients' requests will be known as *service providers*.

A main argument promoting the idea of pervasive computing is that computing power is monetarily cheap, as the most powerful workstations of yesteryear (e.g., late 1990s to 2000) can be bought for a few hundred US dollars. These machines of yesteryear are plentiful and can easily be set up as surrogates. While this is an easy argument to make about the potential computing power of a pervasive environment, one must also consider the number of users in this environment as well as the demand placed on a surrogate by each user. In an urban or densely populated area, the number of mobile devices requesting services may greatly outnumber the available surrogate service providers, and the service rate of the providers may not meet the needs of the users. Thus, the service providers may form the bottleneck in a pervasive environment. The surrogate machines offering the services can be thought of as a public commodity; therefore, it may be difficult to predict their load, and hence, difficult to provision these resources.

While the cost to provision these resources as well as the efficiency of these resources may not be the foremost goals of pervasive computing, there is reason to address these concerns; the bottleneck may degrade the performance to a degree where a user is annoyed by the high latency to perform a task. An element of transparency is therefore lost, as the user becomes aware of the under-provisioned and over-utilized system. Yet, provisioning the proper number of resources may not be possible and the users may have to settle for the best-effort provisioning.

Therefore, under best-effort provisioning, we will examine how resources can be best allocated in order to minimize distraction to the user. However, this resource allocation must not require users to micromanage their own resources, as this should be done transparently from the user in a pervasive environment. Yet, this transparent allocation must also provide users with the same level of satisfaction as if the users were managing their own resources. This section will describe techniques to transparently manage resources at the mobile-host level, as well as techniques to alleviate the bottleneck at the service providers.

2.1   Distributed Caching

One technique to alleviate the bottleneck and congestion at the surrogate level is to distribute the workload of the surrogates among the mobile devices. This is not a novel idea, as its application can be seen in decentralized peer-to-peer (P2P) file-sharing systems. In a decentralized P2P system, the peers act as both clients and servers, as they request and service queries. Regarding file sharing, a peer sends a request for a particular file, and any peer who has a copy of this file can respond to this request by sending the file to the requesting client. Thus, there is no central server to handle requests, but rather, the clients themselves service each other's requests. Therefore, the workload is distributed among the clients instead of being concentrated at a single source, which would act as a bottleneck.

Regarding a service-oriented pervasive environment, a user sends a request for a common service from his or her mobile device to a remote server which handles the request and provides the mobile client with the response. Common services could be a request about the current weather condition or a request about the current traffic pattern. In a centralized approach, all mobile clients would send these requests to the same service provider. A provider would process a request by calculating the most current state of the weather or traffic and relay this response to the requesting client.

Distributing the load in a service-oriented pervasive environment is subtly different than distributing the load in a P2P file-sharing environment. Files in a P2P system are static, and as such, they do not grow stale. The same file can satisfy two requests occurring at two different points in time. Whereas the data requested in a service-oriented system is dynamic. Two requests for the weather at different points in time (e.g., a few days apart) cannot be satisfied by the same response. Therefore, one must keep in mind the idea of freshness of the data in a service-oriented environment. For some services (e.g., weather and traffic requests), the service providers are the only ones capable of producing fresh responses to the queries, as these machines are connected to the weather centre or news centre via a local network or the Internet. A mobile client, on the other hand, only has knowledge of the responses to its own queries as well as those responses to the queries it overhears. It may overhear a query if it is promiscuously listening to the wireless medium, or if the client is being used to forward a query–response pair on the path between the client and server. For other services (e.g., time of day requests), a mobile client may be capable of servicing fresh data.

Thus, while a central server can be eliminated in the P2P file-sharing model since the peers are capable of independently performing a service equivalent to that of the central server, a central server cannot be eliminated in the service-oriented model. The dynamic nature of the data in a service-oriented model suggests that clients will depend on a true service provider to obtain fresh data. A client providing a cached response to a service request is not performing the equivalent service of the true source, since the cache response grows stale over time. Thus, a pervasive environment must incorporate true service providers and cannot distribute all the workload to the clients themselves. A hybrid model is more appropriate, in which the true service providers are used to service the clients demanding the freshest data, and the clients with cached copies are used to service fellow clients who can tolerate stale data. The following section details the advantages of this hybrid caching model.

2.1.1 Advantages of Caching in a Pervasive Environment

The most obvious advantages of employing caching in a pervasive environment are that the bottleneck at the service provider can be alleviated and the response time for service requests can be decreased. The bottleneck is alleviated as requests are dispersed away from the source and directed to the clients. Alleviating the bottleneck lowers the congestion at the source and also lowers the source's workload, which in turn reduces the response time of a query to the source. More significantly though, response time of a query can be lowered if a mobile client chooses to send its request to another mobile device servicing cached data instead of sending this request to the source. A mobile device serving cached data will be most likely be less congested than the actual source and will also be geographically closer than this source. Thus, response time is decreased due to the reduced congestion and a lower round-trip time. However, the most significant advantage of caching is its scalability. As more clients are introduced to the pervasive environment, these clients soon become servers of cached data. Therefore, even though the demand for services will increase due to an increase in the number of clients, the availability of cached data will also increase, and thus, the increase in demand will be met by an increase in supply. One caveat when the number of clients are increased is that as more clients exchange cached data, stale data is propagated to more nodes and the lifetime of this stale data is increased. Stale data propagation will be addressed later in this article, but first, a simple software architecture for implementing a caching system in a pervasive environment will be presented.

2.1.2 Software Architecture of a Caching System

In the protocol presented in [2], mobile devices act as both clients and servers, requesting and responding to service requests. The software architecture of the protocol consists of three entities: Providers, Consumers and Information Managers. All three entities are housed on a user's mobile device. A Provider process simply offers a service, which it can provide to remote devices or to the local device on which the Provider process is located. A Consumer process requests a service. Lastly, the Information Manager is a process that manages all the Providers and Consumers within a single mobile device and also communicates with the Information Managers of remote devices. Figure 1 illustrates the communication between these entities.

In order to provide a service, a Provider will register itself with the local Information Manager. The Information Manager will then advertise this service to all mobile devices within its transmission range (i.e., one-hop neighbours) by broadcasting an advertisement. When querying for a service, a Consumer will send its query to the local Information Manager of the device. The local Information Manager will first check a cache of queries, and if it has a cached response for the very same query, it will respond with the proper answer. Otherwise, it will check the list of Providers that are registered within this device in an attempt to satisfy this query locally. If no local Provider is found within the device, then the query will be broadcast to one-hop neighbours with the hope that a neighbouring device will have a cached response to this query or will have a Provider who can answer the query. In [2], a broadcast query is limited to one-hop as a means to regulate the flooding. As an alternative, one-hop neighbours could subsequently broadcast the query further. However, broadcasting a query aimlessly can easily congest the network, thus it is important to limit the size of the flood. Rather than expanding the flood radius, a host whose query cannot be answered by neighbouring hosts may, instead, send its query directly to a known data source, such as a weather centre when attempting to obtain the current weather conditions.

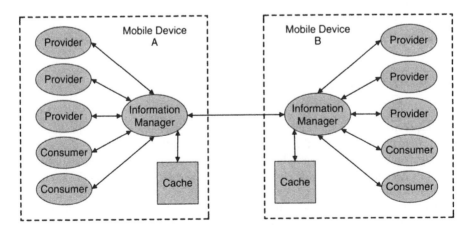

FIG. 1. Communication scheme between Providers, Consumers and Information Managers.

2.1.3   Propagation of Stale Data

While having much in common with a decentralized peer-to-peer file-sharing protocol, the aforementioned protocol [2] also resembles a proxy cache through its use of cached responses. A proxy cache, or web cache, is located between the true web server and a querying client. It can answer a client's request by serving the client a cached HTML page, instead of forwarding the query to the server. In the protocol offered in [2], each Information Manager can cache queries and responses which it overhears or forwards. If the Information Manager receives a query for which it has cached a response, it will respond with this answer. As with a web cache, the response to this query may be stale. In this situation, a trade-off has been made to serve stale data in an attempt to shift some of the load off the service providers. In [2], simple techniques to deal with stale data are offered. First, cached responses have a timeout period, such that an entry is deleted after some period of time. Similarly, instead of deleting an entry after a timeout, the Information Manager may re-query for this service, as a means to update its cache with more current data. Unfortunately, a re-query may result in the same stale data being sent by another remote device, and thus there may be prolonged existence of the stale data in the system. These timeout and update schemes are simple, but do not offer the user much control, nor do they guarantee that fresh data will be received after a timeout. Implicit to the pervasive goal, a user's mobile device should be an unobtrusive tool, and as such, it must be elegantly customized and tuned to its user. A user may require more than a simple timeout to govern the updating of data. The user may want a certain level of freshness when accepting cached data, and therefore, the user may wish to know when the cached data was generated by the source. On the basis of the time at which the data was generated, the user may wish to know the probability that the cached data is fresh. The following section offers a different technique to combat stale data, in which the user controls the level of freshness of the cached data that he or she receives.

2.1.4   Imposing a Quality-of-Service Metric on Cached Data Objects

The previous sections have discussed the advantages of employing a caching system to alleviate the bottleneck at the service providers; nevertheless, the issue of stale data propagation remains a problem. In general, this issue concerning the spread of stale or fresh data is known as data consistency. Some environments require a strong consistency for cached data. For example, in a multi-processor environment in which each processor has its own cache, the same data object may reside in multiple caches; yet when any processor reads this object, this read must reflect the latest write to this object. The requirement where any read must reflect the most recent write is known as strong consistency, and thus, stale data will never be read. In a multi-processor setting, maintainenance of this condition is important to ensure the correctness of an executed program. Relaxing this condition to a point where the latest read of a data object may not return the value of the latest write yields a weaker consistency. The caching protocol discussed in Section 2.1.2 assumed a weak consistency for data objects, allowing the propagation of stale data. The problem of stale data can be addressed by a technique described in [7]. While the offered technique was intended for Internet caching, it is just as appropriate in the domain of pervasive computing, as the service-oriented architecture of pervasive computing is similar to the service-oriented architecture of most websites. This technique allows the user to adjust the level of consistency for the requested data objects in order to meet one's personal needs.

It should be noted that maintaining a strict coherency between a data object at the source and all cached copies of the object held at the clients is very time-consuming. The classical cache coherency protocols come in two patterns: update and invalidate. These require some service to manage a directory, which stores the list of locations of all cached copies as well as state information describing the current access permissions of the clients. In a pervasive environment, the number of clients will likely be too large to manage such a directory. In addition, broadcasting (or multi-casting) the invalidate messages, or the larger update messages, would congest the network with great deal of overhead traffic. Moreover, connectivity to the mobile hosts in a pervasive environment is not guaranteed. Thus, the invalidate and update messages are not guaranteed to reach all of the hosts, and therefore, the coherency protocol would only be best-effort. Settling for a controlled weak consistency is therefore an appropriate solution in a pervasive setting.

2.1.5 Trading-off Consistency for Response Time

When servicing the requests in a pervasive environment where caching is permissible, there is a trade-off between the consistency of the data served and the response time. If a user requires strong consistency, then the request must go to the source serving the data in order to ensure that this user receives the most recent copy of the data (e.g., only the server in the weather centre can provide the most recent weather conditions). If we reasonably assume that there is congestion at the main source, or that the mobile client has intermittent connectivity to the source, the client will experience a high response time when querying this source. On the other hand, if a client can tolerate a stale data object, then a cached copy of the object can be received from one of many less-loaded, and possibly physically nearer, clients, thereby, reducing the response time. In the technique introduced in [7], the user can decide, for each data object, whether consistency or response time is important.

FIG. 2. Quality-of-service domain.

The decision for consistency or response time is not a binary decision, but rather, there is a continuous domain from which a user can select a point between strongest consistency (high response time) and weakest consistency (low response time) for a particular data object. In [7], this domain is known as a quality-of-service (QoS) domain, and it is illustrated in Fig.2. Values in this domain represent the minimum probability, between zero and one, that the data object received will reflect the latest write to this object. By selecting a value of one, the user is guaranteeing with 100% probability that he or she receives the latest version of a data object. By selecting a value of zero, the user is suggesting that he or she is willing to receive any version of the requested data object, no matter whether it is stale or fresh. By selecting an intermediate value, say 0.3, the user is suggesting that he or she is willing to receive a version of the data object where there is a 30% or greater probability that the received object reflects the most recent version.

Users in a pervasive environment can set a QoS value for each service performed by their mobile device. For example, on the one hand, an investment banker would select a high QoS value for the service updating stock quotes. Only the most current quote is important to the banker. On the other hand, a casual investor may set a low QoS value for the stock quote service. Similarly, a golfer checking the weather at the course may set a low QoS value for the weather service. The golfer may believe that the weather does not change drastically within a day, and therefore a stale cached value would be sufficient for offering the approximate temperature and weather condition. Yet, a newscaster reporting on a tornado may require the most accurate weather condition and therefore set a high QoS value for the weather service. Thus, a user must be able to customize this setting for each service.

How can the system determine the probability that a cached object is fresh and reflects the latest write? This probability depends on how often an object is updated and on the time of the last update. To implement this, the source for a service would maintain a table of entries, in which an entry is inserted after an update is made to the data object. An entry to the table consists of the time between updates (the time from the last update to the most recent update). An average can be calculated across these entries to acquire the mean time between updates (MTBU) along with its standard deviation ($STDV_{MTBU}$). The source would maintain the MTBU, $STDV_{MTBU}$ and time-of-last-update values and would provide these to a client requesting its service.

A client will use these metadata values to determine the probability that cached object is fresh and has not been modified. When receiving a request for a cached data object, the servicing client can calculate the probability that the data object has been changed since the time the object was cached. This assumes that the time-between-update values follow a normal distribution pattern. The servicing client needs to first calculate how much time has passed since the time-of-last-update. Using this value along with the MTBU and $STDV_{MTBU}$, the servicing client can calculate the area under the normally distributed bell curve for this object (e.g., by integrating over the probability density function or using a simplified statistical z-score table) to determine the probability that the object has been modified (P_M). Equation (1) describes the probability density function (PDF) for the time between updates, where the independent variable t represents the time since the last update. Figure 3 graphically illustrates this curve. Once the P_M has been determined, the simple calculation in Equation (2) will determine the probability that the object has not been modified (P_{NM}).

$$PDF = \frac{1}{\sqrt{2\pi \cdot STDV_{MTBU}^2}} \cdot e^{\frac{-(t-MTBU)^2}{2 \cdot STDV_{MTBU}^2}} \tag{1}$$

$$P_{NM} = 1 - P_M \tag{2}$$

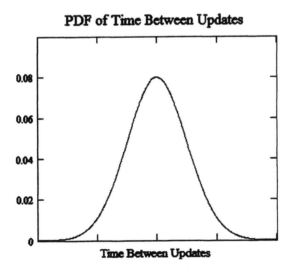

PDF of Time Between Updates

FIG. 3. PDF for the time between updates.

Before a request is satisfied, the requesting client should compare the P_{NM} value of a cached object to the QoS value set for this object. If the QoS value is less than or equal to the P_{NM}, then the request can be satisfied by the cached data. Otherwise, a different servicing client can be sought, or the true source of the object can be queried. It should be noted that the servicing client and requesting client may be one and the same. Before a client uses an object in its own cache, it should calculate the P_{NM} for the object and compare it with its own QoS value for the object.

2.1.6 Alternative QoS Metrics

While the proposed method [7] achieves its goal by allowing users to customize the quality-of-service level of their requested data objects, there are alternative metrics for quantifying *quality of service* which may be more appropriate in a pervasive setting. The offered QoS metrics should give the pervasive user control and flexibility, such that the user attains high satisfaction and minimal distraction. The method discussed in Section 2.1.5 uses freshness as the metric for calculating a QoS value. For services reporting data such as stock quotes or breaking news, the freshness metric is well suited for characterizing quality of service. The changes made to these types of data are important and worth noting. Those individuals requesting these services expect up-to-the-minute data, and thus, the current metric is fitting. However, for a service such as the weather, the freshness metric might not be as fitting. A weather centre may update the current temperature, pressure and wind speed every minute or even every few seconds. Yet, most people do not require an up-to-the-minute report on the weather. Thus, stale data is permissible here. This raises the following question: how should a user set the QoS value for this service? Consider that the user does not mind receiving stale data, as long as the data was generated within the last twelve hours. It would be difficult for the user to set an appropriate QoS value for the object, if QoS values are based on the probability of receiving fresh data. Even once an appropriate probability value is found to satisfy this case, any change in the mean time between updates or standard deviation will lead to change in the range of acceptable objects.

Therefore, the time an object was last modified may be a better quality-of-service metric for some data objects; or an even more complicated metric could be devised based on the payload of the data object. Again referring back to the weather example, an individual typically only cares about significant fluctuations in the temperature. Similar to the mean time between updates, a weather centre could record the mean change in temperature and a user could select a QoS value based on this value. On the days where temperature greatly increases or decreases, a user would require more current data, and on the days where temperature is stagnant a user could accept older data. Even within a single day, temperature may go through periods of change and periods of consistency; for example, the temperature may greatly increase during the morning, remain stagnant during the afternoon, fall during the evening, and remain stagnant overnight. Thus, even though the weather centre may update their temperature reading every minute, a user may only require updates when the temperature significantly changes (i.e., during the morning and evening periods). Therefore, to offer flexibility and customization options to the users in a pervasive environment, multiple QoS metrics should be available.

2.1.7 A Cache Replacement Policy

Query resolution based on the quality-of-service of cache contents brings about the issue of a cache replacement policy. In [7], a technique is offered for replacing objects in a client's cache. This policy attempts to keep the most appropriate objects in the client's cache under the assumption that coherence is not maintained between the cached object and the master copy. The issue of coherence changes the definition of an *appropriate object for caching* between the context of a traditional system and that of a pervasive computing environment. In a traditional system, coherence is maintained, and therefore, when describing an appropriate object, spatial and temporal localities are well-suited indicators of appropriateness. Traditionally, the least recently used (LRU) policy is employed to take advantage of spatial and temporal localities of objects. In a pervasive environment, spatial and temporal localities may still be relevant, yet the opportunity for objects to grow stale should be considered in the replacement policy as well. Objects that are frequently read by mobile clients and infrequently updated by the service provider should be kept in cache, as copies of these objects will remain fresh for a long period of time, and moreover, a client with a cached copy becomes a suitable server for fresh data. On the other hand, those objects that are infrequently read by clients and frequently updated by the provider should be replaced, as these cache copies become stale quickly, and clients with cached copies become poor servers for fresh data.

To create a cache replacement policy based on the previous definition of appropriateness in the pervasive environment, a second set of statistical values needs to be maintained regarding the read history of a data object in order to find the mean time between reads (MTBR). Using the MTBU and MTBR values, a Caching Quality Factor can be calculated, as illustrated by Equation (3). This quality factor is a simple ratio of the MTBU and MTBR. Each cached object will be ranked by this factor; the higher the rank, the more appropriate an object is for caching. Thus, if a new object is to be cached, it must have a higher Caching Quality Factor than the object currently in the cache with the lowest Caching Quality Factor.

$$Caching\ Quality\ Factor = \frac{\text{MTBU}}{\text{MTBR}} \tag{3}$$

2.1.8 Alternative Cache Replacement Policy

The cache replacement policy outlined in Section 2.1.7 keeps the freshest data objects in cache, and as such, mobile clients become effective servers of cached data to their fellow clients; however, clients in a pervasive environment should not be concerned with serving others, but rather, be concerned with serving themselves. The proposed policy [7] is intended for use in a web cache, where the purpose of the cache is to benefit the global population of clients. A pervasive environment is slightly different in that the client itself is hosting the cache in order to serve itself first and other clients second. Users in a pervasive environment have the independence to manage their cache according to their wish. It can be assumed that users are selfish, and therefore, they want a policy whereby a maximal number of their local requests can be satisfied from their local cache. Assuming that users only accept objects that meet the aforementioned QoS setting, it can be shown that the proposed cache replacement policy is not optimal for each user. For example, if a user gives a QoS value of zero to an object which he or she frequently reads, the user would not want the object to leave the local cache. Therefore, even if this object has the lowest Caching Quality Factor, it should not be replaced.

A better replacement policy would consider the number of reads one makes to each object (as in the LRU policy), as well as the QoS value for each object, in addition to the MTBU statistics. Using all these factors, a user can calculate if a cached object meets his or her personal QoS setting. If an object meets the QoS value and this object is frequently read by the user, then it should be kept, and otherwise it can be replaced. Equation (4) describes an Alternative Caching Quality Factor in the range of [0, 1], where F_R is the read frequency as a percentage cache reads for this object, P_{NM} is probability that the object has not been modified, and QoS is the quality-of-service setting. As long as P_{NM} is greater than QoS, this factor is positive, indicating that the cached object meets the user's QoS requirement. When P_{NM} is less than QoS, this factor is negative, indicating that the cached object does not meet the user's QoS requirement. The greater this factor, the more appropriate an object is for caching. Those objects which do not meet a user's QoS requirement will have a negative caching factor and will be replaced by objects which meet the user's QoS requirement. F_R scales the factor such that an object meeting the QoS requirement is scaled up by its read frequency (i.e., making it more appropriate for caching, indicative of an LRU policy), and an object failing the QoS requirement is scaled down by its read frequency (i.e., making it less appropriate for caching, since it is read often but fails the QoS requirement).

$$Alternative\ Caching\ Quality\ Factor = F_R \cdot (P_{NM} - QoS) \tag{4}$$

This alternative cache replacement policy is intended to be user-centric, as users in a pervasive environment are independent, and as such, serving oneself is more important than serving others. Thus, the alternative policy favours objects which meet the individual user's QoS requirements and are read frequently by the user. Consider the aforementioned situation where a user sets a low QoS value for an object. The object will remain appropriate for caching as long as the P_{NM} for that object is greater than its QoS. Thus, even when there is a low probability that an object is fresh, it may be appropriate for one's cache; and the fact that this low-probability object may not be suitable to serve other clients is irrelevant to the selfish user. In addition, the selfish user wants to retain in cache those objects which he or she frequently reads; and the access patterns of other users are irrelevant. In summation, users are not altruistic and should not base their caching decision on their ability to benefit others, but rather on their ability to benefit themselves.

2.2   Broadcasting

As with distributed caching, the simple technique of broadcasting can be used to alleviate the bottleneck and congestion at the service providers. The idea of broadcasting is not unique to pervasive computing, nor is it unique to computing in general. During a broadcast, one speaker delivers information to all the listeners in an area. To apply this idea to a pervasive environment, consider a service provider broadcasting a response of a query to all mobile devices within its wireless range. The mobile devices in range could store a response if it may be of need to the user, or may ignore a particular broadcast. To extend this idea further, consider a service-provider broadcasting data that has not been specifically requested, but there is a high probability that this data is in demand by the mobile users. This is the idea behind the broadcasting of analog radio and television. A listener does not request that a radio feed is sent to him or her, but the listener just tunes into the frequency on which the desired station is being broadcast.

The service-oriented environment offered by pervasive computing could benefit from this basic principle of broadcasting. The benefits of broadcasting in a wireless setting have been advocated in [3], [4] and [6]. First, broadcasting scales well to the number of users in a pervasive environment. Consider the example where commuters wish to know the traffic report in their area. Whether there are ten or ten million commuters, the service provider for the traffic report performs the same amount of work when broadcasting this data. Not only does it scale to the number of listeners, but broadcasting can minimize the workload of the service provider, as the service provider only needs to broadcast a data object once in order to transmit it to the listeners, as opposed to responding multiple times to individual queries requesting this identical data object. This same reasoning can be used to argue that broadcasting can minimize the amount of bandwidth consumed. If a broadcast settles more than one potential request for a data object, then it has saved bandwidth by sending this object in one transmission, rather than multiple individual transmissions. Even if a broadcast only settles one potential request, bandwidth has been saved, as an actual request did not have to be sent to the provider. These scenarios show how broadcasting can minimize bandwidth and the workload of the service provider. Yet, if a broadcast data object is not requested by any user, then the broadcast has wasted bandwidth and wasted resources at the service provider. Bandwidth is the targeted resourced in a wireless setting, as wireless bandwidth is much smaller than wired bandwidth, and therefore it must be allocated more efficiently. Another subtle advantage of broadcasting in a wireless setting is that battery-constrained mobile devices can reduce the number of expensive wireless transmissions if they do not need to actively request certain data objects.

2.2.1 Published vs. On-Demand Data Objects

In [4], techniques are presented to maximize the efficiency of broadcasting data objects, and therefore minimize wasted resources. The context for these techniques is a general wireless setting, yet the same ideas can be applied to the pervasive environment. Data objects, or services in the pervasive environment, are partitioned into two categories: published and on-demand. A published object is one that is broadcast without request, and an on-demand object is one that is only transmitted upon the request of a user. There must be an elegant balance of objects between these types. If all objects are broadcast all the time, resources will be wasted from the broadcast of those objects that no user desires. Thus, only the most popular objects should be broadcast to minimize wasted resources. In addition, the bandwidth allotted to published objects cannot be so great as to limit the resources for those on-demand objects, thereby increasing the access time for the on-demand objects. The constraints of the analytical model [4] for allocating resources between the two object types are as follows: minimize the number of transactions and minimize the access time for a data object. The first constraint implies that a provider should publish as many high-demand data objects as possible, in order to eliminate redundant transactions. The second constraint implies that there should be adequate bandwidth allotted to the on-demand requests. Similarly, there should not be too many published objects in order to give each published item a fair share of the published bandwidth. It is apparent that providers benefit from broadcasting via a reduced workload, but users in a pervasive environment also benefit by means of a less congested network due to the elimination of redundant transactions.

In [4], the authors have modelled the access of data objects as a class-based open-form queuing network. Each data object represents a class with its own arrival rate of requests and service rate. Equation (5) shows the expected access time, t, for any data object. It is defined in terms of the arrival rates, λ_i, for each of the n data objects as well as expected access times, $t_{broadcast}$ and $t_{on-demand}$, for each group of objects. There are k objects in the published group and $n - k$ objects in the on-demand group. Equation (6) is a simplified equation for expected access time for broadcast data objects, where S is the size of an object and B_b is the bandwidth allotted to the published, or broadcasted, data objects. It is simplified in that it assumes that all published objects are of the same length, and there are no replicated objects in the broadcast cycle. Thus, the expected access time for published objects is half the cycle time of a broadcast. A more complete equation for the expected access time of published object can be found in [4]. Finally, the expected access time for on-demand objects, $t_{on-demand}$, is described by Equation (7). It is defined in terms of the aggregate arrival rate for on-demand objects, λ_d, and the service rate for on-demand objects, μ_d. The service rate, μ_d, is further defined in terms of the bandwidth allotted to on-demand objects, B_d, the size of a data object, S, and the size of a request, R.

$$t = \sum_{i=1}^{k} \lambda_i \cdot t_{broadcast} + \sum_{i=k+1}^{n} \lambda_i \cdot t_{on-demand} \tag{5}$$

$$t_{broadcast} \approx \frac{k \cdot S}{2 \cdot B_b} \tag{6}$$

$$t_{on-demand} = \frac{1}{\mu_d - \lambda_d} \tag{7}$$

$$\mu_d = \frac{B_d}{S + R} \tag{8}$$

$$\lambda_d = \sum_{i=k+1}^{n} \lambda_i \tag{9}$$

With basic calculus, Equation (5) can be optimized in order to minimize expected access time. The optimized equation provides the optimal allocation of bandwidth for on-demand objects, B_d, and published objects, B_b. Once optimized, the equation can be used to partition a real set of data objects into published and on-demand groups by the following iterative algorithm:

1. All objects are initially categorized as on-demand.
2. The object with the greatest arrival rate of requests (i.e., most demanded) in the set of on-demand objects is moved to the set of published objects.
3. The expected average access time for this configuration of objects is calculated using the optimized equation for expected access time, and the time is compared to some predefined threshold.
4. If the access time is less than the threshold, steps 2–4 are repeated.
5. Once the access time is greater than the threshold, the algorithm stops and the last configuration to satisfy the threshold is used.

To further save bandwidth, one can batch the responses of on-demand requests [4]. This technique suggests that a service provider should not immediately respond to a single on-demand request with a unicast response to the client. Rather, the provider should wait for a small period of time with the hope that one or more identical queries will be received within this time. After this waiting period, a single multi-cast response can be sent to all clients querying for the same data object. This technique increases the access time for the clients by imposing a waiting time. However, bandwidth is saved whenever multiple requests can be satisfied by a single multicast. Therefore, a trade-off exists between access time and bandwidth controlled by the waiting time. As the waiting time increases, there is a greater chance for more requests to be received and more bandwidth can be saved in the multi-cast, at the expense of a greater access time for the pervasive user. This technique is a blend of the on-demand and published realms, as requests are taken in an on-demand fashion, but responses are sent in the broadcast fashion.

2.2.2 Broadcast Cells

In [3] and [4], there is insight into implementing a broadcasting system in a real-world wireless setting. For a real-world implementation, geographical areas should be partitioned into cells, whereby each cell has an access point for receiving on-demand requests and for broadcasting published objects. Each cell should not necessarily publish the same objects, but rather, each cell should publish the objects which are of greatest demand in that geographical area. For example, in an airport, arrival and departure schedules should be broadcast. In a grocery store, the current sale items can be broadcast. Figure 4 illustrates this idea. To notify users of these published objects, each cell must broadcast a *directory*, which describes a schedule for a time-division multiplexing of data objects. A client can use this schedule to identify the

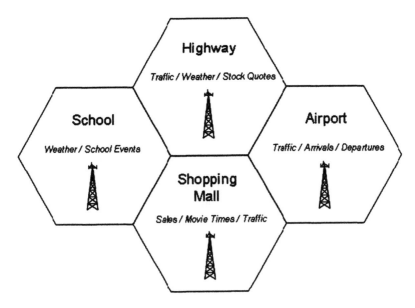

FIG. 4. Partition of a geographical area into broadcast cells (adapted from [4]).

data objects it overhears, as the data objects are broadcast in the order detailed by the directory. When moving between cells, there is no guarantee that the new cell will be broadcasting the same data as the old cell. Even if two cells are broadcasting the same object, the scheduling for this object may be different between the cells; thus for a client to continue to receive a particular service as it moves from one cell to the next, it must read the new directory to determine if the service is broadcast in the new cell and when this service can be retrieved in the broadcast. It should be noted that cells may overlap, and as such, the broadcast of one cell must occur on a different channel(s) than is used by the broadcasts of overlapping cells. The concept of multi-channel broadcasting is described in the following section, and the concept of how overlapping cells can share the wireless medium will become evident.

Further, the idea of partitioning geographic regions into different broadcast cells is appropriate in many real-world applications, as a person's needs are dependent upon his or her geographical location. One's needs while at work in an urban centre are quite different from one's needs at home in a rural suburb; hence the need for location-awareness and location-dependent services varies. To implement this idea, an administrator of a cell could determine which objects are to be published in a cell, or this set of published objects could be dynamically determined. The aforementioned algorithm for deciding which objects are published and which remain as on-demand is applicable to the situation of dynamically determining the set of published objects in a cell. Under the dynamic allocation, all objects would initially be cast as on-demand, and a fixed number of the most requested objects would move to the published group while maintaining some minimal access time threshold. One caveat of the dynamic allocation is the determination of whether an object of the published group should be removed or replaced after some period of time. Once an object is published, the arrival rate of requests for this object will be low, as only those clients who have not read the directory and are not aware of the published objects would send an on-demand request for a broadcast object. Thus, it is difficult to determine how desirable each published object is to the users when the arrival-rate-of-requests metric cannot be used. Yet, even in the midst of this caveat, it should be obvious that dynamic determination of the set of published object is the best solution to meet the needs of users, as these needs vary region by region, day by day, and hour by hour.

2.2.3 Balancing Response Time and Power Consumption

After decisions have been made as to which data objects should be broadcast in a region and how the airtime should be partitioned into broadcast and on-demand time, there are additional subtle issues which need to be addressed, such as how clients locate requested data objects on a broadcast medium and the ordering by which a client retrieves multiple broadcast data objects. These issues can be discussed in the context of a single-channel or multi-channel broadcast. However, before these issues are discussed, there are two goals which should be kept in mind when considering solutions to these issues.

First, the response time for retrieving data objects should be minimized. This helps maintain the distraction-free environment for the pervasive user. Second, the amount of power consumed by the mobile host should be minimized. A mobile host has a limited battery life, and therefore by reducing power consumption, the lifetime of a mobile host is increased. The mobile host's ability to switch between different modes of operation allows it to save power. In active mode, the host can actively listen to the wireless medium, whereas in doze mode, the host cannot listen to the medium as its wireless access card is not in use. Therefore, the ideal way for a mobile host to minimize power consumption is for it to only listen to the wireless medium when it is receiving data objects that it desires. During the time when undesirable objects are being broadcast, the mobile host should not listen and should go into doze mode. Since the benefits of pervasive computing are lost when a user's mobile device(s) is powerless, battery conservation is essential. In addition, the frequent need to recharge one's mobile device creates distraction, which disturbs the seamless integration of technology with one's life. In [6], techniques to address these issues of response time and power consumption are discussed in the context of broadcasting.

2.2.4 Indexing

The first issue which will be discussed is known as indexing. Previously, it was mentioned that a directory was to be broadcast along with data objects to describe the ordering of data objects on the broadcast channel. An indexing scheme has the same objective: inform the clients as to when a data object will be broadcast. An index for a particular data object may be a hash of certain attributes of the data object, such as a filename or URL. A client desiring a data object will first compute this index by performing a hash of the appropriate attributes. Once computed, this index can be used in a variety of ways in order to locate a data object on the broadcast medium. This section will describe two such schemes for locating data objects: distributed indexing and aggregate indexing. Either indexing scheme will benefit the pervasive user by reducing the amount of power consumed by one's mobile device, therefore extending the battery life of the device. The point has already been made that one's mobile device is his or her connection to the pervasive computing world, and thus, it is essential that these devices remain powered for a sufficient period of time without recharging.

Under a simple indexing scheme, known as distributed indexing, an object's index is broadcast immediately before its associated data object. To retrieve a desired data object, a client will listen to all the indices on the medium and when it hears the index that matches that of the desired object, the client will retrieve the data object following its index. What are the implications of this in respect to the two aforementioned goals? Without any indexing scheme, the only items that are broadcast on the medium are the data objects themselves. With distributed indexing, index objects are broadcast along with data objects, and therefore, the length of the broadcast has increased. This is illustrated graphically by Figs. 5(i) and 5(ii). A longer broadcast length implies a longer response time. The trade-off for a longer response time is that indexing allows for a reduction in the power consumption of mobile devices. A mobile client only needs to listen to these short indices and the desired data objects; thus when undesired objects are being broadcast, the client can save power by switching into doze mode.

As an alternative indexing scheme, the individual indices for each data object can be combined into an aggregate index. The aggregate index can be organized as a serial list of individual indices, describing the order of broadcast data objects much like a directory. The aggregate index can also be structured as a tree, which can be searched faster than a list. The organization of the individual indices in the aggregate index may not be of critical importance and will not be discussed further in this

i. No Indexing

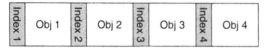

ii. Distributed Indexing

iii. Once-per-cycle Aggregate Indexing

iv. (1, m) Aggregate Indexing where m = 2

v. (1, m) Aggregate Indexing where m = 4

Fɪɢ. 5. Graphical representation of indexing schemes.

chapter; yet it is important to note that aggregate indexing can further reduce power consumption with respect to the reduction achieved by distributed indexing. Under the distributed indexing scheme, a client has to frequently tune into the medium to hear each individual index, switching in and out of listening mode. Switching modes require some power. By aggregating the indices, a client only needs to switch into listening mode once in order to hear the entire index; thus, consuming less power.

Next, when should the aggregate index be broadcast? A simple solution is to broadcast this index once at the beginning of each broadcast cycle (see Fig. 5(iii)). Although under this solution the broadcast length is the same as it is under distributing indexing, the average response time is worse. If a client has not yet read the aggregate index, the client will have to wait half of the broadcast length, on average, before reading the index. Then, the client will have to wait an average of half the broadcast length, again, to retrieve the desired data object. With distributed indexing, a client only has to wait half the broadcast length for the desired index and data object (since these are always broadcast in sequence), but this decrease in response time is at the expense of an increase in power consumption. Another option is to broadcast the entire aggregate index m times throughout the broadcast cycle; this is known as $(1, m)$ indexing. Figures 5(iv) and 5(v) illustrate this indexing scheme with two different values for m. Here, the average waiting time for the index is $L / (2m)$, where L is the broadcast length and m is the number of index replicas in the broadcast. The average waiting time for the data object will again be half the broadcast length. This solution increases the broadcast length, but reduces the response time in comparison to the once-per-cycle broadcast of the aggregate index. Power consumption is the same as the once-per-cycle broadcast scheme as well. Thus, the $(1, m)$ indexing scheme is recommended in [6]. The $(1, m)$ indexing is suitable for a pervasive environment, since less power is required under this scheme in comparison to distributed indexing scheme. Response time under $(1, m)$ indexing is only slightly worse than the alternatives.

Thus, when choosing $(1, m)$ indexing as the appropriate indexing scheme, we are assuming that it is more important to the users to extend the lifetime of their mobile device than to experience a slightly lower response time when retrieving broadcasted data. Due to the nature of the service-oriented architecture in a pervasive environment, response time for broadcasted data does not seem to be critical. The mobile host is expected to predict the user's intent and to retrieve data that it believes the user will want in the near future; thus, data will be ready for the user earlier than if the user were to personally request such data. That is to say, users will not be actively waiting for completion of services, due to the proactive nature of the mobile device in a pervasive setting.

2.2.5 Broadcasting Over Multiple Channels

Another issue brought about in [6] which needs to be resolved is whether to broadcast all the data objects on a single channel, or to distribute the data objects across multiple channels. Until this point, a single broadcast channel has been assumed. The trade-offs of broadcasting over multiple channels will now be explored. The most obvious gain from switching from a single channel to multiple parallel channels is that the length of the broadcast cycle decreases. Assuming d data objects and c channels, each channel must only broadcast d/c objects. Consequently, a shorter broadcast cycle implies a lower response time. Thus, as more channels are introduced, response time decreases. This relationship is true when response time is used to describe the time taken to satisfy a single request. Thus far, response time and power consumption have been analysed under this single-request scenario. Analysis of the impact of multiple channels with respect to these metrics becomes more complicated in a situation where a client is requesting more than one data object. We must note that in a multiple-channel, multiple-request environment, response time describes the time taken to satisfy all requests.

The main issue when expanding to an environment with multiple channels and multiple requests is the possibility for conflicts. It is assumed that a mobile host can only listen to one channel at a time and that switching of channels requires some amount of time and energy. For simplicity, we can assume that each channel is partitioned into the same number of equally sized time slots. Therefore, if two desired data objects are broadcast during the same time slot on two different channels, a client will have to retrieve one object during the first cycle, switch channels and retrieve the other object during the next broadcast cycle. Even if objects are found in adjacent time slots but on different channels (e.g., $Object_1$ is broadcast on $Channel_A$ in slot t_i, and $Object_2$ is broadcast on $Channel_B$ in slot t_{i+1}), a client will have to wait for a second broadcast cycle before retrieving the second object. This is due to the fact that channel switching requires some amount of time.

How do channel switching and conflicts affect response time and power consumption? By nature, switching of channels consumes power; therefore, power consumption increases linearly as the number of channel switches increases. Response time is not so much a function of the number of channel switches, but more a function of the number of conflicts. Yet, it is not as easy to classify the relationship between response time and the number of conflicts. It is obvious that when a conflict is introduced in a situation where there are only two requested data objects, response time will increase because a second broadcast cycle will be necessary. Yet, when conflicts are introduced in a setting where more data objects are being requested, additional cycles may not be the consequence. Figure 6 illustrates these situations. Fig. 6(i) shows how an additional broadcast is necessary when there is a conflict between the only two requested data objects. Figure 6(ii) illustrates a more complicated retrieval where multiple data objects are requested, and two passes are necessary for the retrieval. Figure 6(iii) is an extension of the example from Fig. 6(ii), with an additional sixth object in the retrieval. The sixth object is in conflict with the first data object retrieved in the first pass. However, this sixth object can be retrieved during the second pass, and therefore no additional passes were necessary to resolve this conflict. The response time between 6(ii) and 6(iii) remains the same, even though a conflict has been added. In general, conflicts increase response time, but this does not hold in all cases.

It was shown as to how indexing could reduce power consumption at the expense of a slightly longer response time in the case of a single broadcast channel; however, in the case of multiple broadcast channels, indexing will lower both response time and power consumption. Thus, it is obvious that indexing is beneficial to the pervasive clients under a multi-channel broadcast. First, consider the case where indexing is not used. A client must scan all broadcast channels sequentially to find the desired

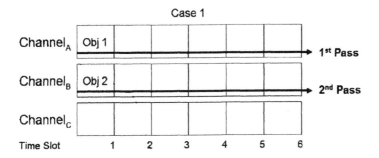

i. The conflict between Objects 1 and 2 results in two passes to retrieve both objects

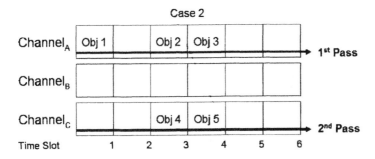

ii. Objects 1 – 5 retrieved in two passes

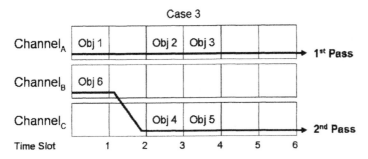

iii. The additional conflict between Objects 1 and 6 does not result in an additional pass to retrieve all objects

FIG. 6. Examples illustrating how conflicts affect response time.

items. If the broadcast length is L time units and there are C channels, the full scan requires LC time units. When indexing is used, the length of the broadcast cycle will increase, but a full scan may be avoided. Consider the simple case where only one object is requested. A client may read an aggregate index from $Channel_A$ and skip to $Channel_D$ to read the desired object in the same broadcast cycle. Thus, scans through the intermediate channels were avoided and response time is lower than it would be under a situation where indexing was not used.

Therefore, for implementation in a pervasive environment, it is beneficial to implement multiple broadcast channels along with an indexing scheme in order to lower the response time for service requests as well as reduce the power consumption of the mobile device. However, the number of channels implemented in a real-word pervasive environment depends on the resource availability. For a wireless medium, the number of channels depends on the number of available radio frequencies. Analogously, the bandwidth of a fiber network depends on the number of fiber cables laid. It is obviously advantageous to increase bandwidth, yet the amount of fiber laid depends on financial resources. This relates back to the motivating factor for this section on resource management. Even if computing power and technological resources are cheap, they are not endless; hence, resource management techniques are necessary. To the point of broadcast channels, a maximal number of channels should be implemented with respect to financial and physical constraints. Once the number of channels has been determined, a $(1, m)$ indexing scheme can be implemented on each channel to describe the broadcast objects. As an alternative, one or more broadcast channels can be solely dedicated to broadcasting indices, while the other channels broadcast only data. Under this solution, the waiting time for an index is reduced, and hence, response time for a data object is reduced at the expense of a dedicated index channel(s).

2.2.6   Retrieval Algorithms

This section will describe three algorithms for retrieving data objects on a multiple-channel broadcast medium. While the details of these algorithms can be found in [6], this section will only present the basic idea of each algorithm in order to find a fitting algorithm for a pervasive environment. With each algorithm, it is assumed that the client has read the aggregate index and knows the broadcast location of each desired data object. Knowing these locations, the retrieval algorithms attempt to find an optimal order by which to retrieve the objects.

The first, and simplest, algorithm is a *Row Scan*, where the term *row* is synonymous with *channel*. Here, a client tunes into each channel, on which desirable objects reside, for one pass of the broadcast cycle. During the pass, the client will retrieve each desired object that is broadcast on the current channel. After one pass, the client

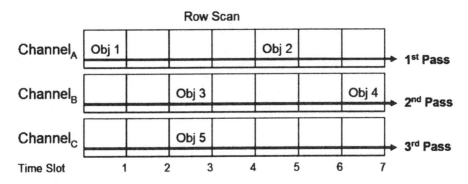

i. Objects 1 – 5 are retrieved in three passes, and two channel switches

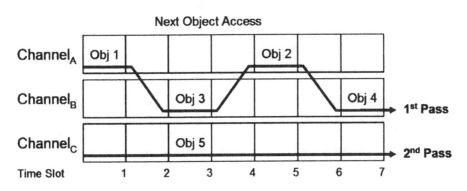

ii. Objects 1 – 5 are retrieved in two passes, and four channel switches

Fig. 7. Row scan and next object access retrieval algorithms.

will then switch to the next channel on which desirable objects reside. Figure 7(i) illustrates this algorithm. The Row Scan provides the client with a minimal number of channel switches. It also requires minimal computation to determine the ordering; a client simply uses the index to determine which channels are broadcasting desirable objects, and sequentially cycles between these channels.

Second, there is a greedy algorithm known as *Next Object Access*. Under this algorithm, a client begins by retrieving the first available object with respect to time. It then continues to select the next earliest object which can be retrieved, no matter in which channel that next object resides. Figure 7(ii) illustrates this algorithm. As a greedy heuristic, it does not guarantee optimal performance. This algorithm will

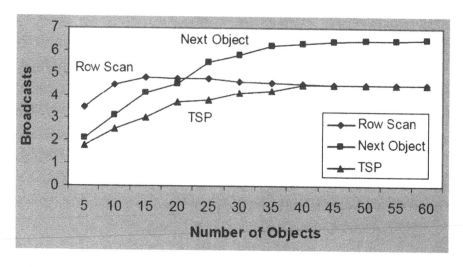

FIG. 8. Simulation results from [6] illustrating how response time varies with the number of requested objects.

usually require more channel switches than a Row Scan, and at best, it will require as many channel switches as the Row Scan. Thus, power consumption is greater under this algorithm. Simulation results from [6], shown in Fig. 8, find that this algorithm provides a lower response time than Row Scan when the number of requested object is low and yields a higher response time than Row Scan as the number of requested object increases. Figure 8 plots response time as the number of broadcast cycles needed to retrieve all requested objects.

Finally, finding the optimal retrieval order of objects in a broadcast is similar to finding the optimal order of cities to be visited in the classical traveling salesman problem (TSP). The goal of TSP solutions is to avoid an exhaustive search through all possible orderings of data objects. Simulation results, shown in Fig. 8, show that TSP heuristics yield a lower response time than both the Row Scan and Next Object Access algorithms. Furthermore, the number of channel switches under the TSP algorithm will be greater than or equal to the number of channel switches of the Row Scan algorithm; therefore, power consumption will not be improved compared with the Row Scan.

Which retrieval algorithm is most appropriate for a client in a pervasive environment? Because users have different needs, the easy solution is to push this decision to the individual user. A user could select some point between minimal response time and minimal power consumption, and the most appropriate algorithm can be chosen. This would be similar to the aforementioned technique for choosing QoS levels for cached data objects, presented in Section 2.1.5. However, regarding the retrieval of broadcast data, the users in a pervasive environment would most likely value power consumption and computational cost over response time. As argued in Section 2.2.4, since a mobile device predicts a user's intent and request services in advance, the response time of these services is not crucial. Reducing power consumption, on the other hand, is important to extend the battery life of a mobile device. In addition, computational cost is important to the pervasive user. Given that a mobile device independently performs services for its user, it is safe to assume that many services and decision-making processes will be competing for computational time on a user's mobile device. Thus, the choice for a power-conserving, computationally simple algorithm which sacrifices response time may be appropriate. The Row Scan algorithm fits this description because it has the lowest computational cost and power consumption and yields a response time that is close to that attained by the TSP heuristics, especially when the number of requested objects is high.

2.3 Adaptive Fidelity

Another resource management technique which can help create a distraction-free user environment is fidelity adaptation. The term fidelity is used to describe the quality of an offered service; thus, fidelity adaptation implies a dynamic adjustment in the quality of a service in order to meet the current resource constraints. Lower fidelities require fewer resources, while higher fidelities require more resources. Examples of fidelity parameters include frame rate and resolution for a streaming video service. In this example, a streaming video player may lower the frame rate or resolution if bandwidth is decreased due to a congested network. If the bandwidth increases, the frame rate and resolution can be increased. For traditional services, users can control these fidelity parameters to meet their needs. A user may request low-resolution video stream if his or her bandwidth is low, or may request a web page without images in the same scenario to reduce the latency. In a pervasive environment, a user should not be bothered about manually setting fidelity parameters each time resource levels change. Micromanagement of fidelities takes away from the transparent user experience, in which technology is seamlessly incorporated into the everyday activities of the user. Thus, the system must make dynamic decisions about fidelity without user input. There are two questions which the system must answer when making decisions on fidelity adaptation: How do different settings of fidelity parameters affect resource consumption? What are the user's preferences in terms of which fidelity factors to adjust in a given situation? Section 2.3.1 will address the first question by presenting a technique to determine resource consumption as a function of fidelity parameters. Section 2.3.2 will address the second question by presenting a technique that determines the level of user satisfaction (i.e., utility) provided by a configuration of fidelity parameters.

2.3.1   Modelling Resource Consumption as a Function of Fidelity

Narayanan et al. have designed an empirical method for estimating resource consumption as a function of fidelity. This method captures and logs resource usage levels as well as the associated fidelities during the run of an application. On the basis of these historical values, resource consumption can be modelled as a function of fidelity at the expense of computation and storage overhead. As an alternative to this technique, an analytical model can be used, whereby the application developers provide a function for resource consumption in terms of the fidelity factors that their application offers. Yet, it would be difficult for developers to construct a function that is generic enough to apply to a variety of machines with different hardware configurations. In order to tailor such a generic function to any hardware configuration, the function would be to incorporate a sufficient number of input parameters that can be used to describe the architecture and organization of any machine (e.g., processor speed and memory size are two such input parameters). Unfortunately, systems are defined by more parameters than simply by clock rate and memory size. Other important parameters include memory organization, processor organization, processor architecture, cache hit rate and many more. These parameters all affect how resources are consumed, but the incorporation of these parameters into a function is difficult. However, there is no need for the empirical model in [5] to consider any of these parameters, as it only uses historical statistics as its basis for estimating resource consumption. Thus, its simplicity makes it a better choice than the complicated analytical approach. The empirical approach can be applied to any application on any machine.

The empirical method implemented in [5] is broken into three phases: logging, learning and an online phase. During the logging phase, a service is run at different fidelity levels while the resource consumption is monitored by hardware and software monitors. The learning phase uses the logged results to build an estimation function for resource consumption in terms of multiple fidelity factors. For an inexpensive computation, a linear regression is used to map the fidelity parameters (e.g., frame rate and resolution) to the consumption level of a single resource (e.g., computation time or bandwidth). At the end of the learning phase, each resource is estimated by its own function. Initially, the logging and learning phases are performed offline. Thus, when a user first uses a service, the logging and learning phases will have been performed, such that during the online phase, the fidelity parameters of the service can be tuned to meet the resource constraints. In order to tune these parameters, the current resource constraints (e.g., available CPU and bandwidth) become the input parameters to the functions, and the maximum fidelity levels that satisfy these constraints are returned.

Equations (10) through (13) show an example of the set of functions determined by the learning phase. In this example, the consumption of four resources is modelled by four functions with respect to three fidelity parameters. During the online phase, the availability of each resource will be monitored and will become the input to these functions. The functions will be used to determine the appropriate levels of fidelity to satisfy these resource constraints. Consider the case where $Resource_1$ represents bandwidth, and there is 10 Mbps of available bandwidth. Equation (10) will then be fixed to 10 Mbps and solved to determine the configurations of the three parameters (i.e., $Param_1$, $Param_2$, $Param_3$) that can satisfy the constraint of 10 Mbps. Consider that these parameters are frame rate, resolution and audio quality. There will be multiple configurations of frame rates, resolutions and audio qualities that satisfy the bandwidth of 10 Mbps. Which configuration is most appropriate? As the available bandwidth decreases, should frame rate, resolution or audio quality degrade? In [5], it is assumed that there exists a utility function that details a user's preference between fidelity parameters. Hence, the utility function may show that a user prefers a drop in frame rate to a drop in resolution as bandwidth is reduced. The following section will take a deeper look into techniques to estimate the utility functions of a user.

$$Resource_1 = f_1(Param_1, Param_2, Param_3) \tag{10}$$

$$Resource_2 = f_2(Param_1, Param_2, Param_3) \tag{11}$$

$$Resource_3 = f_3(Param_1, Param_2, Param_3) \tag{12}$$

$$Resource_4 = f_4(Param_1, Param_2, Param_3) \tag{13}$$

2.3.2 Modelling Utility as a Function of Fidelity

As part of Project Aura [9] at Carnegie Mellon University, an analytical model has been developed to determine the configuration of fidelity parameters that maximizes a user's utility in the presence of resource constraints [8]. The analytical model efficiently finds an optimal configuration of fidelity parameters for a service by considering a set of service providers, a set of fidelity configurations for each provider and the utility provided to the user by each configuration. For example, fidelity configurations for a video player would be the pairs of frame rates and resolutions that satisfy a given bandwidth. It is assumed that given a fixed resource level, it is possible to find all the configurations of fidelity parameters which satisfy the resource constraint. The functions constructed by the empirical technique presented in Section 2.3.1 can provide such configurations. Once this set of configurations is known, the configuration which provides the user with the greatest utility should be chosen.

The proposed model calculates a single utility value for each configuration of fidelity parameters. Utility assumes a value in the range of [0, 1], where zero implies that a user is completely unsatisfied, while one implies that the user is completely satisfied. The simplest way to map utility values onto fidelity configurations is to have the user choose a utility value for every possible configuration. However, the number of possible configurations can be undoubtedly large. For example, the set of configurations for a streaming video player is the Cartesian product of the frame rate and resolution offerings. Consider that the set of offered frame rates includes 20, 30 and 40 frames per second (FPS) and the set of resolutions includes the qualitative values *high* and *low*. The Cartesian product of these two fidelity parameters produces the complete configuration domain shown in Table I. This domain grows quickly when either the number of fidelity parameters is increased (e.g., adding a third fidelity parameter of audio quality) or the set of offered values for a particular fidelity parameter is increased (e.g., adding 50 FPS to the set of offered frame rates). To avoid this large configuration domain and its rapid growth, each fidelity parameter can be considered independent from one another. Therefore, instead of mapping a utility value to each configuration in complete configuration domain, a utility value only needs to be mapped to each offering in the individual fidelity-parameter domains. In our example, instead of mapping utility values to each of the six possible configurations in the complete configuration domain, utility values only need to be mapped to each of the three frame rate offering and to each of the two resolutions.

For discrete fidelity-parameter domains, such as the ones presented in Table I, a mapping table suffices to capture the mapping between utility and fidelity. These mapping tables would need to be manually set by the user in a pervasive environment. These could be set once in an initial offline setup procedure. The user would not be distracted by such configuration decisions after the initial setup. For continuous domains, such as the volume of an audio track, a mapping table cannot be used since there are an infinite number of elements in these continuous domains. Instead, a sigmoid function can be constructed to capture the relationship between utility and fidelity over a continuous domain. Figure 9 depicts a utility function in the form of a sigmoid function. The function asymptotically approaches a lower limit and an upper limit. By knowing the range and the knees of a sigmoid function, a continuous

TABLE I
DISCRETE FIDELITY-PARAMETER DOMAINS

Frame Rate Domain	{20, 30, 40}
Resolution Domain	{*high*, *low*}
Complete Configuration Domain	{(20, *low*), (30, *low*), (40, *low*), (20, *high*), (30, *high*),(40, *high*)}

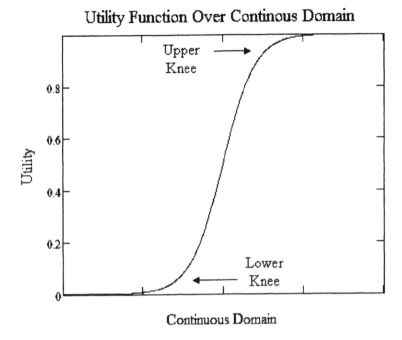

FIG. 9. Utility function over a continuous domain.

function can be interpolated. Therefore, storing a utility function for a continuous domain parameter only requires storing the two knee values, as the upper and lower limits are known to be one and zero, respectively. Upon setup, a user needs to manually determine the knees. These can be prompted to the user by asking him or her for the domain value which is insufficient for the service (lower knee) and a domain value which is good enough for the service (upper knee).

Equation (14) is the basic formula for determining the maximum utility over a set of service providers and fidelity configurations within each provider. The inner most term, $F_p^{w_p}(c_p)$, is the aforementioned utility function for an independent fidelity parameter. This function is either given by a mapping table or a sigmoid function. The power term, w_p, is a weight that denotes the importance of a fidelity parameter to the user. This is also a manually set parameter in the range of $[0, 1]$, where zero is the highest weight and one is the lowest. This weighted utility value is calculated for each fidelity parameter in a configuration and the results are multiplied together; the resulting product will be between $[0, 1]$. This result denotes a user's overall utility from a single configuration offered by a service provider. A consequence of multiplying the individual utility values together to form the overall utility is that if the utility received by a single parameter in the configuration is zero (i.e., completely insufficient), then the overall utility will be zero and will indicate a completely insufficient configuration. Finally, this product is then multiplied by the value, F_s, indicating a user's preference for the service provider. Again, this value is manually set by the user in the range of [0, 1], where zero denotes that the provider is insufficient and one denotes that the provider is completely sufficient. A user may prefer one service provider over another due to some qualitative features of the providers which are not captured by the utility function. For example, Emacs and Microsoft Word are both word processors, but a user may prefer Microsoft Word over Emacs because of the formatting options offered by Microsoft Word. The value, F_s, allows a user to express these preferences.

$$Maximum\ Utility = \max \left[F_s \cdot \left(\prod_{p \in QoS\ Param(s)} F_p^{W_p}(c_p) \right) \right] \tag{14}$$

When finding the maximum utility value, the work of an exhaustive search through the configuration space can be reduced by implementing an elegant stop condition. First, service providers are to be visited in descending order of their F_s values. A utility value for each configuration is calculated and the greatest value is saved as the current maximum. Before repeating this process for the next service provider, the F_s value of the next service provider is compared with the current maximum utility value. If the next F_s value is less than the current maximum, the algorithm can be stopped, as the current maximum will be the global maximum. This is true because the aggregate product term is upper-bounded by one, and therefore the overall utility is upper-bounded by F_s for a service provider. Therefore, a service provider does not need to be considered if its F_s value is below the current maximum. By visiting the providers in descending order, it is safe to stop the algorithm once this condition is met.

The proposed algorithm achieves its goal, but it has its limitations. The analytical model offers a solution to find the configuration of fidelity parameters that offers the highest utility. It is distraction-free during the online phase, but requires the user to set: (i) utility values for the elements in the individual fidelity domains; (ii) weights for each fidelity parameter; and (iii) preferences for each service provider. It makes the assumption that by considering each fidelity parameter independently, the optimal global configuration will be found. Finally, its early termination condition can only be used if a user has different preferences towards service providers. When this condition is used, the algorithm will only show a great improvement over an exhaustive search if there is significant deviation in the preferences towards service providers. When homogeneous service providers are considered, those for which a user has equal preference, an exhaustive search must be performed. Therefore, while the analytical approach may reduce the storage by assuming independence of fidelity parameters, it has not been proven that fidelity parameters should be interpreted independently. Furthermore, the search time can only be reduced when there is enough deviation between a user's preferences towards service providers.

2.4 Conclusions

This section covered three solutions to the resource management problem in pervasive computing: distributed caching, broadcasting and adaptive fidelity. One motivating factor for these solutions comes from the fact that pervasive computing is intended to be service-oriented, and thus there is foreseeable congestion at the service providers. Therefore, to prevent congestion at the service providers, the solutions must be scalable with respect to the number of clients which are sending requests to these providers. The techniques of distributed caching and broadcasting were offered to alleviate this congestion in a scalable manner. A second motivating factor for these solutions comes from the fact that many of the clients in a pervasive environment will be mobile devices, whose resource capacity is low. Resources such as battery power, wireless bandwidth, CPU time and memory must be used efficiently in order for the mobile device to provide the user with a distraction-free experience. Realizing that the independent actions (i.e., service requests) made by a mobile device may place a high demand on system resources, a technique was offered to adaptively adjust the fidelity of the services in a manner that provides the user with the most utility. Resource conservation, specifically regarding power consumption, on the mobile device was also considered when a choice was made for the appropriate implementation for a broadcasting system in a pervasive environment.

Distributed caching shows its scalability, which is obvious from the fact that clients become service providers for cached data. As the number of clients increases, the number of service providers increases as well. However, the caveat is that cached data becomes stale over time. To combat stale data, a technique was presented to allow users to select some quality-of-service level for the data objects they receive. For some data object, a user may sacrifice response time to obtain the most recent version of the object, or may relax the constraint for fresh data, thereby decreasing the response time. The mean time between updates was the key metric in defining quality of service.

Broadcasting proves its scalability, which is obvious from the fact that as the number of clients receiving broadcast data increases, the resources required to broadcast this data remains the same, since the service provider does not process requests for this data, nor does the provider need to transmit more data onto the network as the number of clients increases. An algorithm was presented to partition the bandwidth between broadcast and on-demand data objects, in order to minimize the access time of requests. Furthermore, the idea of indexing was presented as a solution to minimize power consumption at the clients, with the only trade-off being an increased response time under a single-channel allocation. As the number of channels increases, response time decreases. It is necessary to implement a proper retrieval algorithm to acquire all desired data objects in the face of conflicts. In a pervasive environment, the choice for a retrieval algorithm as well as for an indexing scheme depends on the trade-off between power consumption and response time. By slightly sacrificing the optimal response time, indexing can be implemented to greatly decrease power consumption, and the Row Scan retrieval can be employed to offer the minimal number of channel switches. Since one's mobile device is intended to issue requests without user interaction, response time is not crucial, for the user will not be actively waiting for each issued request. The effect of power consumption on the battery life of the mobile device is important, with the goal being to prolong the usage time of the device. Thus, a small increase in response time is worth a larger decrease in power consumption.

Finally, the adaptive fidelity technique is capable of effectively managing the resources of a user's mobile device. In a service-oriented pervasive environment, it is foreseeable that many applications will compete for the limited resources on the mobile devices, resulting in high utilization, which will degrade the performance of the individual applications. The presented technique for adaptive fidelity monitors the resource usage and adjusts different fidelity factors of each application according to the current resource constraints and predefined user preferences. When a resource such as bandwidth becomes constrained, applications will lower their dependence on this resource by adjusting appropriate fidelity factors, such as the resolution of received images. By predefining preferences to each fidelity factor, the user will not be distracted with on-the-fly decisions to increase or decrease the load on resources in order to better the performance; these decisions will be made behind the scenes, transparent to the user. This solution is fitting for a pervasive environment, since it prevents the user from being distracted by an over-utilized system and also shifts the decision-making responsibility of adjusting fidelity factors away from the user and onto the system.

3. Security

The concept of an environment saturated with computing agents that seamlessly interact with one another as well as with human users lends itself to many potential exploitation hazards. Security measures will be required in order to guarantee the integrity of operations; however, the authentication overhead must be limited to ensure seamless interaction. Two main identification issues have been recognized and addressed in the literature [11]:

1. **User identification:** the computing environment needs to be aware of the identity of each user in order to properly handle access to data and services. The smoothness of the identification protocols is particularly critical when synchronizing and transferring data across multiple platforms.
2. **Service identification:** users need a method to check that a particular service is trustworthy before engaging in its use. This prevents malicious services, which may be disguised as trusted services, from infecting a mobile client with malware or viruses. It also gives the user a sense of comfort and confidence when calling a service, such that the intended service is the one that is actually called.

Although they are both related to the common area of minimal-overhead security maintenance, these two tasks are very different in nature and need separate treatment. We will address strategies for user and service identification in the next two sections.

3.1 User Identification

The field of user identification is one of the unresolved issues in the context of computing services, and many secure solutions have been elaborated. First, it is appropriate to distinguish between user *authentication*, where the user claims an identity and the system verifies the validity of that claim, and user *recognition*, where the system proactively identifies a user (usually based on physical features) without any action on the user's part. User authentication can be further divided into two categories: *knowledge-based* (i.e., the identity claim is validated on the basis of some information provided by the user, usually a password or a PIN) and *token-based* (i.e., the identity is confirmed through the possession of a token, which can be anything from digital keys to smart cards).

Quite obviously, user recognition is conceptually more appropriate to the pervasive idea of a smart environment than user authentication, as the user will not need to undertake any distracting action to gain access to desired services. However, recognition is computationally more challenging than authentication, in that the system will have to compare the user's features against a potentially large database of identities in order to determine if the presented features match any stored entry. Authentication, on the other hand, only requires the comparison of presented information (i.e., username and password) against a single database entry in order to determine if the presented password matches that of the stored password for the presented username.

This distinction implies that the relative efficiency of authentication with respect to recognition can only be determined on a case-by-case basis, establishing which of the two (the user-initiated process of authentication, or the waiting time to search a database in recognition) is more bothersome for the user. In the next two sections, we will examine two user identification methods: token-based authentication and biometric recognition. We will compare and contrast the two approaches and describe the respective benefits and limitations.

3.1.1 Token-based User Authentication: Kerberos Protocol with Smartcards

In the effort to strike an optimal balance between low computational complexity (and therefore transparency) and security of operations, secure hardware has promising features. It provides secure storage for highly sensitive information (such as the user's cryptographic keys) and also offers the ability to perform cryptographic operations in hardware, thereby allowing for efficient and transparent authentication strategies. The user is not responsible for the initiation and execution of authentication protocols, since the secure hardware contains all the necessary information to complete the authentication in an independent manner.

One possible strategy that has been proposed involves the integration of Kerberos V5 with a smartcard [10]. In order to understand the role played by the secure hardware, let us first examine an ordinary Kerberos authentication:

1. As a preliminary step, all users and services need to have a long-term cryptographic key registered with an Authentication Server.
2. When a registered user wants to login to a particular service, it will send the Authentication Server a request for an initial ticket to connect to the Ticket Granting Service. The Authentication Server will then provide a session key encrypted with user's long-term key.
3. The user will receive the session key, decrypt it and use it to request a ticket from the Ticket Granting Service (TGS) in order to connect to the desired service. The Ticket Granting Service will now return a session key encrypted with the initial Ticket Granting Service key, rather than the user's long-term key.

Kerberos thus provides two layers of indirection in order to limit the use of the user's long-term key, therefore increasing the security of this protocol. Short-term session keys are used to encrypt the communications between the user and service provider. Due to the volume of transactions encrypted with a short-term key, these keys are susceptible to be being discovered by an eavesdropper. Yet, a discovered key only allows the eavesdropper to listen to a single session, as these keys change session by session. The long-term key is used in very few transactions, preserving its security. However, two issues remain unaddressed:

1. The long-term key will have to be used occasionally, since the session keys from the TGS are only temporary. Furthermore, the encryption/decryption operations will be performed on the user's workstation, which is not necessarily a secure place.
2. In order to guarantee a sufficient level of security, the long-term key will have to be protected by a safe password, which the user will have to enter every time the key is required. In a pervasive scenario, this might result in an excessive burden on the user.

In [10], an authentication method is used where the user's long-term key is stored in a smartcard. All the encryptions and decryptions are performed by the card, which guarantees the security of operations (addressing the first problem) and eliminates the need for a user password (addressing the second problem), in that all the information necessary to login is securely stored in the card.

Note that this process will of course still suffer from the drawbacks of token-based authentication methods; above all is the risk connected to the token's misplacement. Whoever possesses the smartcard can authenticate as the smartcard's owner. However, simply introducing PIN-activated smartcards, at the expense of a little user interaction, might ease the problem.

3.1.2 User Recognition: Biometric Applications

The basic working principle of biometric recognition can be summarized as follows:

1. A photometric sensor detects the user and converts the relative information into a digital form.
2. The digital information is then processed by a feature extractor, which isolates the traits which are relevant to the recognition process.
3. The user's features are then compared against a template database. The result of the comparison will return the user's identity or a no-match.

The first step of biometric recognition poses the question: which human characteristics should be used in the detection process? Ideally, these characteristics will have to satisfy two conditions: (i) the trait will have to be unique and non-reproducible, in order to unambiguously identify a user and avoid system circumvention and (ii) the information should be easy to acquire and acceptable on the user's end, since a technology that is invasive (like a retinal pattern read) or has criminal reminiscences (such as fingerprints) may be unwelcome to users and may disrupt an otherwise seamless environment.

There are currently several biometric sensors on the market and under development. Biometric sensors can be used to recognize facial features (using optical or thermal techniques), as well as a human iris, written signatures, fingerprints and hand geometry. Facial thermography proves to be perfect solution, for it satisfies the two aforementioned requirements, as it is resistant to forgery and is least invasive. A thermographic sensor will detect and map the temperature distribution on the user's face. Since the blood flow pattern seems to be unique to each individual (and also hardly modifiable through surgery), the identification is highly secure. The features are also relatively easy to acquire.

The second and third steps of biometric recognition bring about a different aspect of identification, namely, the computational cost and performance. As mentioned earlier, one of the fundamental requirements of a pervasive environment is the transparency of operations; not only should tasks be performed with minimal (if any) user intervention, but their execution should also be as invisible as possible. The process of matching biometric features against a template should therefore carry a computational complexity that does not exceed a small fraction of the system's resources.

As a last remark, note that security and invisibility are two conflicting objectives, in that increasing one will necessarily degrade the other. This is well expressed in Fig. 10, which shows the False Match Rate (FMR) – False Nonmatch Rate (FNR) plane. FMR represents the probability of erroneously detecting a match (i.e., when the matching procedure is not restrictive enough, due to computational limitations such as the pervasive requirement for invisibility), while FNR represents the probability of failing to detect a match (i.e., when the matching procedure is too conservative, as in situations where security is of utmost importance).

Different applications will result in different trade-offs between FMR (computational simplicity) and FNR (security). A high-security access application will have very restrictive matching rules, whereas applications where a match is missing is undesirable (like criminal identification and forensics) and will have a high FMR at the expense of FNR. Determination of a pervasive environment's collocation on the graph is an open problem that requires careful examination of the computing capabilities of the system, the sensitivity of data and applications in the system's domain, and the details of the identification process (complexity of feature extractor, size of user template etc.).

When compared to the solution provided by secure hardware, biometric recognition offers a solution that is more transparent and resistant to forgery. At least for the type of identification described in this section, the recognition process is completely invisible and hassle-free. However, the present level of the biometric technology is far from the ideal, exploitation-proof condition that is required for a secure application.

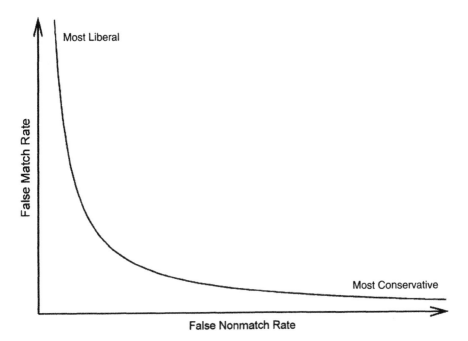

Fig. 10. Optimal set in the False Match Rate–False Nonmatch Rate plane. These two objectives, which represent, respectively, the computational simplicity and the security associated with a recognition method, are conflicting, and cannot be maximized simultaneously. An appropriate trade-off between FMR and FNR depend on the goals and requirements of each specific application (image adapted from [11]).

Feature extractors will also play a fundamental role in determining how secure and computationally expensive the process will be.

On the other hand, smartcards are not associated with any of the previous issues related to pattern recognition. The identification is performed in the traditional, well-established cryptographic key methods, which are not associated with FMR and FNR issues. However, the use of smartcards is less invisible than biometric recognition and will always be subject to the risk of misplaced or stolen cards, even in the case of password-protected cards. While the risk of false identification with biometric systems is likely to be reduced by the constant improving technology, the risk of false impersonation due to fraudulent card possession will be very hard to control.

3.2 Service Identification

A similar, yet distinct, security issue that is inherent to pervasive systems is the integrity of the services available to the user. In the pervasive scenario, the user does not initiate jobs; instead the jobs are proactively launched based on user preferences, history and context. Therefore, it makes sense to provide service identification in addition to user identification. In other words, the system will have to be endowed with an authenticating structure that filters out possible fraudulent services and restricts job initiation capabilities and data access only to those services that have been authorized.

Similar issues are already present in the area of network systems, where users are constantly faced with the task of locating services (such as printing or storage services), verifying the trustworthiness of these services and submitting a query to those trusted services. Many different methods for service lookup and discovery have been proposed in the past [13] [15], with philosophies and designs that are fitting to the pervasive idea. Such architectures usually involve the existence of discovery and authentication servers, which mediate the interaction between users and services by providing information about service capability, service authentication and user authentication. In most cases, they will invoke the participation of other actors, such as additional certificate databases that physically hold the required information. The structure helps users search for particular services and verify their identity. At the same time, services will broadcast their presence and capabilities within the network, with the possibility of filtering (or customizing) the set of amenities offered to each user.

Similar types of network architecture, presently used to manage and secure the use of distributed computational resources in the standard sense (users initiating processes), can finally be adapted to pervasive system by letting the services, rather than the users, control the execution of tasks. The missing link necessary for the last step is represented by the integration of intelligent devices with the above architectures, i.e., the capability for storing the user's profile and performing pattern recognition on the present context. In the following section, we will discuss the secure Ninja Service Discovery Service (SDS) architecture [13].

3.2.1   Secure SDS with Ninja

The Ninja architecture is based on five classes of agents:

1. The users (clients)
2. The services
3. The SDS servers
4. The Capability Managers
5. The Certificate Authorities

Each SDS server governs its own domain, which comprised a number of agents from the other four classes. Domains are organized in a hierarchical fashion, allowing for easy scalability in case of overload; whenever a SDS server is unable to handle all the services and broadcasts in its domain, it will start a new child SDS server that will take over a sub domain. The following shows the interaction between the main entities:

- Each SDS server sends authenticated messages over a global multi-cast channel, which includes a description of the domain itself, descriptions of services registered in the domain, the multi-cast group address for each service in the domain, the address of the Certificate Authority and the address of Capability Manager.
- Each service is responsible for registering itself with a live SDS server (who multi-casts their presence via periodic advertisements). This implies that when a SDS server crashes, a service under this server is responsible for registering itself with a new server; this ensures some degree of fault tolerance. A service periodically multi-casts its service descriptions (using authenticated and encrypted messages) to its SDS server as well as clients in its multi-cast group.
- Each client submits its queries for services to the SDS server responsible for its domain. The server will then reply with a list of matches corresponding to the client's query, the available resources and the user's privileges. It is assumed that there is trust between the clients and the SDS server, such that the client can trust that the service descriptions received by the SDS server are accurate. However, this does *not* imply anything about the functionality or correctness of the services; it simply verifies the descriptions of the registered services. The client can then choose a service from list and join the multi-cast group on which this service is being transmitted.

The core of the security framework of SDS is contained in the encryption of all messages between the system's entities, especially between servers and services. The use of asymmetric encryption would be the best choice for all encryptions, but efficiency requirements (which play an important role in simple network systems, and even more so in pervasive environments) suggest that a hybrid symmetric/asymmetric method would be best. Therefore, the service multi-casts follow the three-segment format illustrated in Fig. 11. The first part of the message contains the sender ID. The second part, ciphered with the SDS server's public key,

| Sender ID | Ciphered Text (containing symmetric key) | Payload (encrypted with symmetric key) |

FIG. 11. The format of a service broadcast.

contains several pieces of information (again the sender ID, the destination, etc.) along with a symmetric key that can be used to decipher the third, and largest, portion of the message, which is the actual payload. Thus, computationally expensive public-key decryption is only necessary to obtain the symmetric key, while computationally cheaper symmetric-key decryption can be done on the larger payload. This reduces the decryption overhead while at the same time securing the messages against eavesdropping.

In addition to encryption, the system implements a global authentication procedure to guarantee the integrity of the associations between the system components and their public keys. In other words, security against fraudulent identities must be guaranteed not only by encrypted communications but also through authentication of the endpoints. This is the role of the Certificate Authority and is accomplished in two steps:

1. The Certificate Authority collects certificates from the various system components.
2. Clients can query the Certificate Authority for a certificate to assess the validity of a public key associated with a service.

Since the keys and the certificates are public, the service of a Certificate Authority would not require computationally expensive encryptions when in operation (as the Certificate Authority only performs the signing of certificates offline) and therefore would blend well with a pervasive system where thin agents with low latency responses are a priority. The only requirement for implementing a Certificate Authority is that it must reside on a secure server.

The last component in this architecture is the Capability Manager, which stores the lists of clients' privileges in order to determine which user has access to which services. This greatly simplifies the amount of user interaction needed for each single query, since the SDS servers will prompt the Capability Manager for possible access restrictions, instead of asking the user to authenticate. In addition, the user is only returned a list of matches which he or she is authorized to use; all other services are effectively invisible to the user.

Let us consider the example structure in Fig. 12 depicting an SDS that manages resources in a computer science building. The hierarchy comprises four SDS servers, which can either be responsible for the administration of a specific physical location (such as the 'CS Hall', '4th Floor' and 'Room 443' servers) or for the control of certain services (like the 'Systems' server). In this figure, solid lines represent one-time communications, while dashed lines represent the system's various periodic broadcasts. One-time communications include the Remote Method Invocations by which each server can generate additional servers, as well as the clients' queries

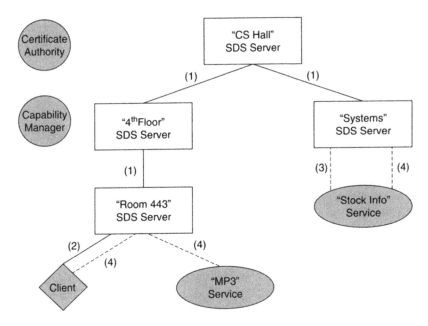

FIG. 12. The structure of a SDS architecture: dashed lines indicate periodic broadcasts, while solid ones represent one-time communications. Lines marked with (1) are the authenticated server connections between a server and its offspring; lines marked with (2) represent the authenticated client connection; lines marked with (3) are the service broadcasts and lines marked with (4) are the server broadcasts (image adapted from [13]).

for services. The periodic broadcasts include the periodic server broadcast that disseminate the information about the server domain and the service broadcasts that publicize the available facilities.

From a pervasive perspective, this architecture provides an infrastructure that satisfies the demand for a transparent, yet trustable, provisioning of services. The identification of services is assigned to the SDS servers, which keep track of their identity (which has been verified by a Certificate Authority) and only share with the clients the existence of those services with trusted identities. All a client needs to do is submit a service query to the server, and the resulting matches will be, by construction, already authenticated. As an aside, the structure provides the additional benefit of a straightforward organization of the user's privileges through the Capability Manager. Lastly, it should be strongly noted that the level of trust guaranteed by this architecture is only as strong as the trust between a client and the Certificate Authority and the trust between a client and the SDS server.

In order to incorporate this architecture into a pervasive system, an additional step is required to eliminate the need for explicit queries for services made by the user, transferring this task to the control of the mobile device itself.

3.3 Conclusions

This section has presented a set of available tools and platforms to address security issues in the specific context of a pervasive environment, i.e. in a system with strong constraints on transparency and minimal user distraction. The issues of identification have been separated into two categories: user identification and service identification. Furthermore, user identification has been partitioned into authentication and recognition, and each has been reviewed in order to offer a spectrum of different (and possibly, complementary) identification mechanisms which realize different degrees of security and transparency.

Finally, we have described a model service identification architecture equipped with a selective discovery protocol that not only prevents unauthorized users from using the available resources but also prevents them from detecting those services to which they have not been granted access, increasing both the efficiency and the security of the system in a context-adaptive fashion.

4. Current Projects

As a realistic illustration of the challenges posed by resource management and security requirements, as well as the practical solutions that have been devised over the years, we present here a few projects that are currently under way.

The applications of pervasive computing span the entire spectrum from generic, multi-purpose distributed computing to the implementation of specific services like home automation or pervasive healthcare. An extensive list of ongoing projects is shown in Table II. This section will focus on a selected group of these projects, including:

- Multi-purpose systems providing proactive services to the users: MIT's Oxygen, Carnegie Mellon's Project Aura and Berkeley University's Smart Dust all belong to this category.
- The Pervasive Continuous Curriculum (PCC) project from Pennsylvania State University [14] and AULA from University of Castilla investigate the application of pervasive systems to academic issues, such as the administration of classes and the design of individual curricula.

TABLE II
A SAMPLE OF CURRENT PERVASIVE COMPUTING PROJECTS BASED
ON CATEGORY

MULTI-PURPOSE SYSTEMS
Oxygen (MIT, [16])
Aura (CMU, [17])
(Berkeley, [18])
Spectacles (Johannes Kepler Universität Linz, [19])
PerComp (Federal University of Campina Grande, [20])
Application SuperNetworking - All-IP (University of Oulu, [21])
Particles (University of Munich, [22])
Mundo (Technische Universität Darmstadt, [23])
MOBIUS (European Mobius consortium, [24])
OTOGI (Waseda University, [25])
ACAMUS (Kyung Hee University, [26])
LOCAL (University of Minho, [27])
Disappearing Computer Initiative ([28])

EDUCATION
PCC (PSU, [14])
AULA-IE (University of Castilla - la Mancha, [29])

HEALTHCARE
MyMD (MIT, [30])
Context Aware Health Monitoring (University of Technology, [31])
Abaris (Georgia Institute of Technology, [32])
TMBP (Denmark Centre for Pervasive Healthcare, [33])
Centre for Pervasive Healthcare, [33])
UniCare (Imperial College, [34])

HOME, OFFICE AND URBAN AUTOMATION
LiveSpaces (University of South Australia, [37])
FlexHaus (Fraunhofer Institut SIT, [38])
SSLab (Keio University, [35])
SmartLab (University of Deusto, [36])
Interactive Workspaces (Stanford University, [39])
Cityware (Imperial College et al., [40])
Shared Worlds (University of Limerick, [41])

- Medical services: the projects MyMD (from MIT) and TMBP (Centre for Pervasive Healthcare, Denmark) provide a framework for healthcare monitoring.

- Smart environments naturally lend themselves to home, office and urban automation applications. Existing projects propose the implementation of pervasive devices at several different scales, spanning from the automation of daily home tasks (like powering on/off the lights or operating appliances) to the concept of connectivity and service provision as integral components of urban design and architecture.

4.1   Multi-Purpose Systems

One of the founding ideas of pervasive systems is the concept of a single user served by a multitude of computing agents, which saturate the environment in order to monitor the context, predict the user's intent and offer a wide range of services (ideally, all services that are necessary to the user and do not require his or her interactive participation) in a proactive fashion. A vast number of current pervasive projects deliver a generic framework for the provision of services, from a simple weather forecast update to more complex business transactions. In this section, we will describe two such frameworks: MIT's Oxygen and Project Aura from Carnegie Mellon University.

4.1.1   Transparent Security with Oxygen

Traditionally, the interaction between humans and computers has required humans to learn and adapt to the logic and working principles of the specific machinery at hand. Conversely, the Oxygen project orbits around the idea of bridging this interaction by teaching computers to communicate in a human-friendly manner, supplying the user with service-providing agents that completely mask the underlying technology. This framework involves three principal entities: Users, Devices and Networks:

- **Users:** the human clients, and main focus, of the system. In other words, the computational platform provides a set of technologies that enable the user to automate tasks, network with other users, and communicate information in a completely natural and transparent fashion:

  o *Automation:* low-level actions are represented by basic automation objects, and user technologies include scripting tools capable of manipulating these objects and constructing arbitrarily sophisticated actions from the basic building blocks. The objects can be *physical* (which can include perceptual devices, temperature or light sensors and power switches among others) or *virtual* (which comprise software agents and daemons capable of processing information and making decisions). As an example, a set of physical objects could be combined in a script specifying user preferences such as indoor temperature, light and sound levels, computer screen resolution, preferred font size and so forth; this script could then be run whenever the user enters a building or a computer lab, allowing him or her to concentrate on high-level tasks rather than on adjusting the environment settings.

o *Collaboration:* the system also provides a platform that keeps track of the interactions between users, using context-aware agents to classify the content, the properties and the parties involved in each specific collaboration instance. This information is then passed to the global system to provide collaboration-related services, such as teleconference infrastructure, scheduling of meetings, and collaboration in database management.

o *Knowledge access:* data can be produced, searched and shared among users with the aid of the knowledge access subsystem, equipped with semantic search capabilities and tools for extensible data representation and acquisition. The Haystack platform [42], the Semantic Web [43] and the START language [44] are all components of this subsystem.

- **Devices:** the entities responsible for detecting the user's intent and providing the appropriate services. They can be portable or embedded in the environment.

  o Users are provided with Handy21s (H21s), i.e. handheld devices that are associated with a specific user, rather than with an environment. The handhelds provide continuous connectivity for the users wherever they are, at any time.

  o Vehicles, buildings and public spaces are assigned one or more computing agents called Enviro21s (E21s). These are responsible for the provision of services that pertain to the given environment, such as receiving and sorting telephone calls within a certain building or delivering travel information and driving directions in a user's car.

A visual representation of these two categories is illustrated in Fig. 13, showing the implementation of devices in a given environment: the space is pervaded by several Enviro21s (for instance, one in each room of a building); in addition, the mobile devices Handy21s move across the domain communicating with the relevant Enviro21s and between each other, as appropriate.

- **Networks:** the infrastructure that establishes connectivity between users and devices. The networks (N21s) can dynamically reconfigure themselves to provide

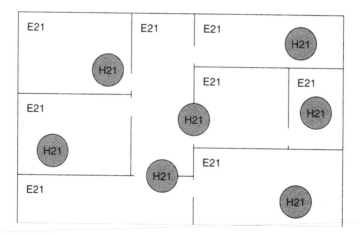

Fɪɢ. 13. A typical oxygen environment, with fixed devices embedded in each section (for instance, each room of a building) for the provision of room-related services such as controlling the appliances and guaranteeing the access to a certain set of users only. Handheld devices move through this structure communicating with the ambient devices and between each other, and negotiating all the low-level procedures that are required by each action, thereby absorbing most sources of user distraction.

the interconnection between subsets of computers, also referred to as *collaborative regions*. Networks also play a fundamental role in the discovery of services by mobile agents. When a user enters a new smart space, the user's handheld device will automatically explore the surrounding environment, through the provided network, and record the available services for future use.

Oxygen thus provides a realization of several of the security techniques described in Section 3. For ordinary operations, the system resorts to a token-based method of authentication. The tokens are, in this case, the handheld devices H21s, which are furnished with the capabilities to authenticate the users with the surrounding services, eliminating the need for any other type of access control (the system is also endowed with a Discovery Service architecture responsible for resolving the service requests). Furthermore, the token-based authentication can be combined with more secure identification methods (such as fingerprint identification implemented on the H21s) for those specific applications that require an increased level of protection, such as bank transactions or access to sensitive information. In addition, the H21s can be instructed to identify each other, in order to form a collaborative region, i.e. a self-organized set of mutually authenticated users who may share data or have access to specific services. The routing infrastructure is provided by Chord [45], a scalable framework for peer-to-peer overlay networks.

4.1.2 Project Aura: Resource Management by a User Proxy

Project Aura from Carnegie Mellon University is a pervasive platform based on the concept of a personal *aura*, i.e. an abstract representation of the user's intent and preferences, which facilitates the user's mobility by taking control of all the migration-related duties [17].

In particular, the project addresses four types of causes of user distraction due to heterogeneous computing infrastructure: (i) a migration to a different environment, as the user moves between different locations and chooses to transfer the pending tasks across devices; (ii) a change in available resources, as network connectivity and/or computing power fluctuation due to time-varying system load; (iii) a change in the executing task, due to a change in the user's task priority; and (iv) a change in the context, which requires the suspension of prior tasks and the inception of new ones.

Ideally, the user proxy plays the role of a coordinating entity that decides on the services to request, on the quality-of-service that can be considered acceptable and on all the other issues related to the aforementioned distraction sources. The system architecture comprises four components:

- The **Task Manager**, or **Prism**, constitutes the user proxy. Prism resorts to a platform independent description of tasks, treated as high-level, conceptual objects rather than just as a collection of specific applications. This allows the system to be more aware of user intent (e.g., denoting one's action as 'editing a text document' as opposed to 'using Microsoft Word') and to easily migrate the tasks between different platforms.

- The **Service Suppliers** provide the service wrappers for the specific environment they reside in. The Suppliers will respond to an abstract service request (such as 'editing a text document') by invoking the specific application (e.g., Microsoft Word) present in the current environment. This is the system component that targets the heterogeneity of computing environments by encapsulating all applications with the required capabilities in one high-level wrapper. The other system components do not need to be aware of these details.

- The **Context Observer** monitors the physical context, passing the relevant information to the other units in charge of the migration of tasks. The Context Observer is, for instance, responsible to detect when a user leaves a specific environment and joins a new one.

- The **Environment Manager** plays the role of a domain coordinator, monitoring the available resources and organizing the users' requests.

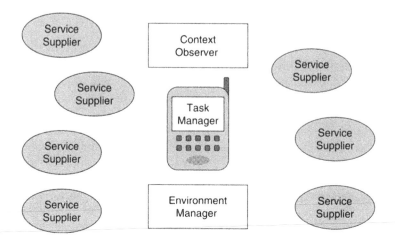

Fɪɢ. 14. The outline of Aura's architecture.

Every environment is equipped with an Environment Manager, a Context Observer and an instance of Task Manager for each user in the environment. The environment also contains Suppliers for each available service, which is registered with the Environment Manager through a XML-based feature description system. A pictorial representation of the system interactions is provided in Fig. 14.

Project Aura represents a solution to the resource management problem that relies on a complex unit (the Prism) making smart decisions based on context and user intent.

4.2 Dedicated Systems

The concept of computing agents that permeate the environment has also stimulated work in the realm of specific applications, such as education, healthcare, unmanned vehicles and home automation.

Quite understandably, predicting user intent and preferences in a dedicated environment is simpler than it would be in a generic environment, as the system architecture can be tailored to the specific properties and constraints of the application at hand. In other words, it is feasible to trade some of the flexibility of multi-purpose systems for simple and accurate prediction of user intent. We will now describe two application-driven scenarios, illustrating how some of the issues of non-specialized frameworks find here a natural solution.

4.2.1 Pervasive Education

The task of curriculum design in collegiate programs is an elaborate, often time-demanding procedure that involves evaluating the student interests, establishing the required *corpus* of knowledge or expertise needed for a specific degree and avoiding, to the largest possible extent, duplication of material across courses.

There is a conspicuous list of issues connected with such a task:

- **Curriculum development and access tools:**

  o To each degree, it is necessary to associate a set of modules that constitute the essential, indispensable core of knowledge for that specific curriculum. This set should also be as uniform as possible across different institutions, in order to guarantee that the same denomination is indeed affixed to the same educational path; this is particularly suited to computer science and engineering degrees, but can be extended to other areas.

  o On top of the standardized core, each student can build a collection of individual subfields of expertise. The selection of the appropriate units involves the interaction between the student, his or her academic advisor, and the faculty members offering course modules in related areas. This process presupposes the existence and availability of relevant, up-to-date information, and the ability to process this information efficiently.

- **Teaching practices:** The course offering itself should present a similar degree of flexibility and 1-1 interaction between parts. In the first place, class syllabi should be fine-grained to allow individual students to only select the topics which they have not yet learned. Ideally, the granularity should eventually be so fine that the degree consists in a continuous knowledge acquisition rather than a discrete course-based learning process. Second, interactive methods should both facilitate the course offering and increase classroom participation, making sure the student's progress is fed back directly into the system.

It is apparent that the advances obtained in the area of database management, learning structures, distributed computing, mobile agents and pervasive systems provide a feasible solution to many of these issues. As an example, the Pervasive Continuous Curriculum (PCC) project constitutes an effort to collect the relevant technologies emerging in these fields in order to construct a pervasive education framework. The platform includes three sets of components:

- The set of instructors, I
- The set of students, S
- The set of courses, C.

The members of each set and their interactions are realized on the backbone provided by the Pervasive Information Community Organization (PICO) framework [14], which consists of software agents (called Intelligent Delegates or *delegents*) that can self-organize into dynamic communities with the purpose of sharing data between one another, processing different sources of information, and making context-aware decisions. For example, as a course is scheduled within an academic program, a course delegent D_c is created, which holds information about the course syllabus and records the student delegents D_s that are created for each student that registers for the course. The D_s will perform extensive checks on each student's background to determine whether he or she meets the prerequisites to attend the class and whether the class contents match the student interests and/or satisfy the chosen degree requirements. The D_s will also interact with the instructor's delegent D_i to create an effective 1-1 learning scenario, where individual questions and difficulties are addressed on an adaptive, personalized basis. An illustration of the system's components and interaction is shown in Fig. 15.

4.2.2 Pervasive Healthcare

In this section, we will describe how a pervasive system can offer continuous healthcare monitoring for patients with critical medical conditions. Quite understandably,

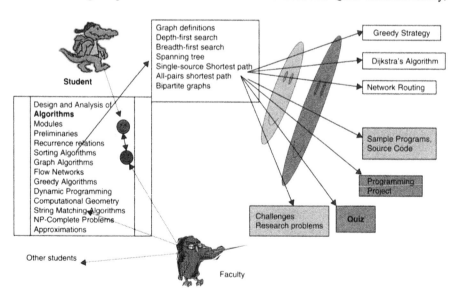

Fɪɢ. 15. A model of the interactions involved in course administration using a computer science course in algorithms as an example.

the delicate issues involved in patient treatment demand for a strict set of constraints regarding the system's quality-of-service and, to some extent, privacy and security.

One such system is MyMD, a project developed at MIT [30]. The platform is composed of five entities:

- Sensors, which monitor the patient's vital signs
- Sensor proxies, in charge of combining and coordinating the readings of all the different sensors
- A real-time streaming database, for fast access to the patient's history (both for reading and updating purposes)
- Storage capabilities, to provide database support
- Communication facilities to promptly issue alerts and provide emergency directions

The realization of this architecture poses several challenges within the domain of resource management: the medical devices will need to be portable yet autonomous entities, capable of performing their task with minimum energy consumption. The system will also have to constantly, proactively monitor the network conditions in order to predict possible failures and promptly effectuate the transition to alternative procedures in case communication is lost.

4.3 Conclusions

In order to provide a realistic sample of the features and requirements for the implementation of a pervasive system, we have reviewed a sample of ongoing pervasive computing projects, along with their strategies to tackle the challenges described in Sections 2 and 3.

A crucial element in this scenario is represented by the following inherent challenge to pervasive systems: while dedicated applications come with intrinsic and well-defined notions of the minimum quality-of-service and security levels necessary for safe operation, generic platforms required by pervasive environments must be more flexible to satisfy the heterogeneous requirements of the clients, service providers and intermediate networks.

5. Final Remarks

Pervasive computing is the third generation of computing in which many computing devices, of different shapes and forms, concurrently serve the individual user. Much of the groundwork for pervasive computing, with respect to hardware, is already present, such as wireless networks, powerful mobile devices and an abundance of workstations and servers. Furthermore, many traditional theories and algorithms are applicable to pervasive computing. Yet, there is an important property of pervasive computing which requires these traditional hardware and software elements to be modified; it is the requirement for a transparent integration between the user and the computing resources. Services must be performed for the user in a least-invasive, distraction-free manner. A possible resource bottleneck at the surrogate servers, who act as service providers to multitudes of mobile devices, may disturb the users' interaction in the pervasive environment. In addition, the constant need to authenticate oneself with different services in the environment can obstruct the user's interaction. This chapter presented ways to smoothen this interaction and to maintain a distraction-free environment.

Regarding resource management, the techniques of distributed caching, broadcasting and adaptive fidelity were offered. Distributed caching and broadcasting alleviate the bottleneck at the service providers. Distributed caching spreads the workload across the mobile devices, while broadcasting eliminates the need to process redundant requests and transmit redundant responses. Both techniques scale well and are therefore suitable for a large-scale pervasive environment. Since mobile clients become mobile servers with distributed caching, as more clients enter an environment, the increased client demand will be met by an increase in server availability. In addition, the spread of stale date in a caching environment can be thwarted by imposing some quality-of-service guarantee on the data received. With regards to broadcasting, the resources consumed in a broadcast are independent of the number of clients, thus implying the scalability of this technique. Moreover, the power of a mobile host can be conserved when an indexing scheme is integrated with a broadcast, and the response time for a request can be decreased as more parallel channels are introduced. Furthermore, to efficiently utilize the limited resources of a mobile device, an adaptive fidelity technique can be implemented to dynamically adjust the fidelity of services in a manner which yields the greatest utility under the given set of resource constraints. The presented techniques for adaptive fidelity are suitable for pervasive computing as they offer portability and limited user distraction. Since resource consumption can be modelled empirically, regardless of the architecture and organization of a machine, this aspect of an adaptive fidelity algorithm can be ported to the diverse assortment of machines in a pervasive environment. By considering fidelity parameters independently from one another with respect to their offered utility, less setup is required from the user and less storage is required for these settings, thus, limiting the interaction from the user and limiting the storage requirement on the resource-constrained mobile device. With sufficient resources in the available service providers as well as in the mobile clients and communication network, these resource management techniques would not be necessary; yet for a realistic implementation of pervasive computing, resource management techniques should be employed to compensate for an under-provisioning of resources and to allow for a distraction-free environment.

We have also discussed how multiple security issues must be addressed before pervasive systems become a trustable environment to perform important tasks. In the first place, the identities of the system's clients need to be verified in a way that is both secure and transparent. While the integrity of the system operation is naturally a non-negligible concern, the pervasive philosophy brings about an additional, conflicting priority: to minimize distraction to the user, allowing his or her focus to remain on the high-level tasks. The design of valid identification strategies that integrate themselves with the seamless ensemble of pervasive components is an extremely complex task. Token-based authentication with smartcards and biometric recognition have been discussed and contrasted against each other in order to present a profile of the complications involved. In addition, the idea of a smart environment initiating processes on behalf of the user demands for an additional identity check: the services and the service providers themselves will have to be authenticated before they can be integrated with the system. We have discussed an architecture that provides service identification without placing any additional burden on the user's workload. Before a similar type of organization becomes effective in the pervasive context, the concept of users querying for services will have to be replaced with the concept of processes initiating services based on the situation and on the user's preferences and history. Drawing upon the field of artificial intelligence to supply the groundwork for context-aware devices will provide this last step towards the third generation of computing.

As a practical counterpart of our analysis, we have reviewed a few current attempts to achieve a safe yet transparent pervasive operation, commenting on the many different strategies and designs.

ACKNOWLEDGMENTS

We wish to thank Robert Collins for providing helpful comments on this work and Kevin Grady for carefully reviewing and revising this manuscript. This work in part has been supported by the National Science Foundation under the contracts IIS-0324835.

REFERENCES

[1] Weiser M., September 1991. The computer for the 21st century, *Scientific American*, **256**(3): pp. 94–104.
[2] Perich F., Joshi A., Finin T., and Yesha Y., May 2004. On data management in pervasive computing environments, *in IEEE Transactions on Knowledge and Data Engineering*, Vol. **16**, No. 5, pp. 621–634.

[3] Imielinski T., Viswanathan S., and Badrinath B. R., May–June 1997. Data on air: organization and access, *in IEEE Transactions on Knowledge and Data Engineering*, Vol. **9**, No. 3, pp. 353–372.

[4] Imielinski T., and Viswanathan S., October 1994. Adaptive wireless information systems, *in Proceedings of the Special Interest Group on DataBase Systems*, Japan, pp. 19–41.

[5] Narayanan D., Flinn J., and Satyanarayanan M., December 2000. Using history to improve mobile application adaptation, *in Proceedings of the Third Workshop on Mobile Computing Systems and Applications*, Monterey, CA, pp. 31–40.

[6] Hurson A. R., Jiao Y., and Shirazi B., 2006. Broadcasting a means to disseminate public data in a wireless environment: issues and solutions, *Advances in Computers*, Vol. **67**, pp. 1–85.

[7] Sustersic J., and Hurson A. R., 2005. Quality of Service (QoS) in Internet Cache Coherence, *Journal of High Performance Computing and Networking*, Vol. **3**, No. 5/6, pp. 296–308.

[8] Poladian V., Sousa J., Garlan D., and Shaw M., May 2004. Dynamic configuration of resource-aware services, *in Proceeding of the 26th International Conference on Software Engineering*, pp. 604–613.

[9] Sousa J., and Garlan D., 2002. Aura: an architectural framework for user mobility in ubiquitous computing *in Proceeding of the Third IEEE/IFIP Conference on Software Architecture*, Vol. **224**, Montreal, pp. 29–43.

[10] Itoi N., and Honeyman P., 1999. Practical security systems with smartcards, *in Proceedings of the Seventh Workshop on Hot Topics in Operating Systems*, Arizona, pp. 185–190.

[11] Jam A., Hong L., and Pankanti S., 2000. Biometric identification, *in Communications of the ACM*, Vol. **43**, Issue 2, pp. 90–98.

[12] Satyanarayanan M., 2001. Pervasive computing: vision and challenges, *in IEEE Personal Communications*, Vol. **8**, Issue 4, pp. 10–17.

[13] Czerwinski S. E., Zhao B. Y., Hodes T. D., Joseph A. D., and Katz R. H., 1999. An architecture for a secure service discovery service, *in Proceedings of the Fifth Annual ACM/IEEE International Conference on Mobile Computing and Networking*, pp. 24–35.

[14] Hurson A. R., Jean E., Ongtang M., Gao X., Jiao Y., and Potok T. E., 2007. Recent advances in mobile agent-oriented applications, *in Mobile Intelligence: When Computational Intelligence meets Mobile Paradigm*, L.T. Yang and A.B. Waluyo (editors), John Wiley & Sons.

[15] Mauro J., and Minden G., 2004. Security model in the ambient computational environment, M.Sc. Thesis, University of Kansas.

[16] Project Oxygen's website: oxygen.csail.mit.edu.

[17] Project Aura's website: www.cs.cmu.edu/Thura; Sousa J., and Garlan D., 2002. Aura: an architectural framework for user mobility in ubiquitous computing environments, *in Proceeding of the Third IEEE/IFIP Conference on Software Architecture*, Vol. **224**, Montreal, pp. 29–43.

[18] SmartDust's website: robotics.eecs.berkeley.edu/pister/SmartDust

[19] Spectacles' website: www.pervasive.jku. at/Research/Projects/SPECTACLES

[20] PerComp's wiki: wiki.percomp.org

[21] All-IP's website: www.mediateam.oulu.fi/projects/allip

[22] Particles' website: particles.teco.edu

[23] www.tk.informatik.tu-darmstadt.de/Forschung/Poster/Mundo

[24] Mobius' website: mobius.inria.fr

[25] Interaction Group's website: www.dcl.info.waseda.ac.jp/groups/intg.html

[26] Acamus' website: uclab.khu.ac.kr/camus

[27] LOCAL's website: get.dsi.uminho.pt/local

[28] DC's website: www.disappearing-computer.net

[29] AULA's website: chico.Inf-cr.uclm.es/AULA_IE

[30] MyMD's website: mymd.csail.mit.edu

[31] www-staff.it.uts.edu.aut~peterl/mobilelab/research/project1_health.html

[32] Abaris' website: home.cc.gatech.edu/julie/24

[33] TMBP's website: www.tmbp.dk

[34] UbiCare's website: www-dse.doc.ic.ac.uk/Projects/ubicare

[35] SSLab website: www.ht.sfc.keio.ac.jp/SSLab

[36] SmartLab's website: www.smartlab.deusto.es

[37] University of South Australia e-World Lab page: e-world.unisa.edu.au

[38] www.sit.fhg de/_SITProjekte/flexhaus

[39] Interactive WorkSpaces' website: iwork.stanford.edu

[40] Cityware's website: www.cityware.org.uk

[41] Shared Worlds' website: www.shared-worlds.org

[42] Haystack's website: haystack.csail.mit.edu

[43] Semantic Web's page: www.w3.org/2001/sw

[44] START's website: start.csail.mit.edu

[45] Chord's website: pdos.csail.mit.edu/chord

Computing with RFID: Drivers, Technology and Implications

GEORGE ROUSSOS

School of Computer Science and Information Systems
Birkbeck College
University of London
Malet Street, London WC1E 7HX, UK

Abstract

Radio Frequency Identification or simply RFID has become an integral part of modern computing. RFID is notable in that it is the first practical technology to tightly couple physical entities and digital information. In this survey, we cater to the computing professional who is not familiar with the specifics of RFID, which we discuss in the context of supply chain management, its most popular application. We begin with a primer on supply chains, with particular reference to the relationship between efficiency and information flow. We recognize universal identification with bar codes and electronic data interchange as the two principle computing technologies that have played a central role in the optimization of supply chains. We then discuss RFID and supporting network technologies and identify their novel features and capabilities. We proceed by examining the performance improvements in supply chain management due to RFID and differentiate between different levels of tagging. We explore consumer applications and services using item-level RFID in particular. Such applications not only offer novel opportunities for business but also raise important social and policy challenges primarily related to privacy protection, which we discuss in more detail. We conclude by exploring how European law is attempting to address the new issues arising from the use of RFID and look ahead at the challenges encountered when computing with RFID before it can be made an effective end-user technology.

ADVANCES IN COMPUTERS, VOL. 73
ISSN: 0065-2458/DOI: 10.1016/S0065-2458(08)00404-X

1. Introduction

Several authors would have you believe that RFID is the greatest information technology innovation: it will deliver cheaper, better quality and safer food for the global market; it will simplify the manufacturing of cars and airplanes; it will save the environment by allowing every single product to be recycled; it will save human lives by preventing medical mistakes; it will make the world a safer place by averting acts of terrorism; it will do away with counterfeiting, especially of drugs; and of course it will spark the next computing revolution by creating the Internet of Things. Few computer technologies have sparked such excitement.

In this survey, we attempt to separate fact from fiction and develop an understanding of RFID based on evidence and the experience gained through field implementations of this technology. A common theme will be RFID as the catalyst for change in business information system implementations due to its capability to intimately link physical and digital assets and establish relationships that can be processed automatically without need for any manual intervention. For this reason, and despite its relative simplicity, RFID has found numerous applications. Its influence is nowhere more pronounced than in the supply chain, where its popularity has been growing rapidly. There are already several very large-scale deployments of RFID within this sector, which often also takes a leading role in the development of RFID technology. RFID in the supply chain and its extensions in consumer services will also be at the centre of our discussions.

We structure this discussion as follows: first we introduce the basics of supply chain management and the role that computing plays within it. Then, we provide an analysis of RFID technology in this context and identify the role that it can play to provide novel information sources that significantly enhance its efficiency. Yet, the use of RFID in the supply chain has unintended consequences, especially when objects are tagged at the item – rather than the container – level. We conclude by reviewing such implication with particular reference to privacy protection and identify areas where law and policy have to play a significant role if RFID would have a long-term effect.

2. Supply Chain Basics

Supply chains are at the core of modern globalized open markets. Each supply chain has unique characteristics and requirements but they all comprise a network of coordinated organizations, which collaborate in diverse activities to transform raw material and components into finished products and deliver them to the end consumer. Such material and information resources move link by link from supplier to retailer across the supply chain, adding value at each stage, bringing the product farther from the point of production and closer to the point of consumption.

A simplified example of a supply chain for grocery products is displayed in Fig. 1: Raw materials are received by suppliers, who process them in usable forms for example, turning polystyrene and polypropylene granules into plastic film rolls

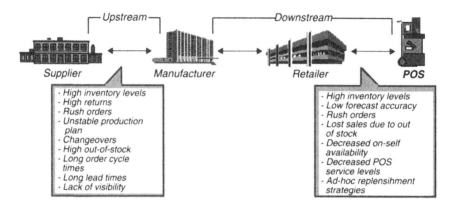

that can be used for packaging, or fresh milk into pasteurized milk and stored in large containers suitable for travel over long distances. Processed materials are received by the manufacturer and are used to fabricate and package the product which is then transported to a retail distribution centre. From this location, products are delivered to retail outlets and displayed at the point of sale for purchase by consumers. To be sure that this is a somewhat simplified view of the process, as at each link there would be more than a single bilateral relationship for example, several suppliers would be needed to provide the full list of materials required for the manufacture of a particular product, and many manufacturers would deliver products to the same distribution centre. Nevertheless, Fig. 1 provides a good model for thinking about the process and helps identify the main issues related to the performance of each step of the process. In practice, the majority of supply chains would include a much longer and complex network of exchanges which spans great distances and more often distances no longer than state borders. Needless to say that supply chains provide great variation. As a point in case, consider the delivery of munitions to the field needed to support operations for the Department of Defense, or the special traceability requirements of so-called cold chains where products are temperature and environment controlled.

One could argue that it is possible to avoid such complexity and the complications of developing and maintaining a multi-partner supply chain by keeping full control of the whole process within a single company. Although this idea may be conceptually attractive, in actual fact this approach would require a single organization of enormous size, which in some cases would far exceed even the largest companies in existence today. In fact, there is some evidence that such a massive organization would be highly inefficient and would suffer due to internal difficulties that would negate any benefits derived from the internalization of the supply chain. Furthermore, collaborative supply chains have gained prominence as a result of the globalization of production and commercial activity and due to the dominance of network effects within this environment. Consequently, supply chain management has increasingly attained greater significance and is today a core factor in establishing competitive advantage. In turn, this fact has brought into focus the role of business relationships which extend beyond traditional enterprise boundaries.

According to the Council of Supply Chain Management Professionals, supply chain management (SCM) encompasses the planning and management of all activities involved in sourcing, procurement, conversion and logistics management. These activities include coordination and collaboration with channel partners, which can be suppliers, intermediaries, third-party service providers and customers. Note that SCM activities travel both upstream (from retailer to supplier) and downstream (from supplier to retailer) across the supply chain. For example, new products (travelling downstream) could be produced and delivered as a result of an order (transmitted upstream) placed to the pertinent distribution centre by a particular outlet.

The principal metric for measuring SCM success is consumer satisfaction, that is, whether the product on offer completely satisfies the needs of a particular consumer and if it is available for purchase at the appropriate time and location as required. This task clearly requires that demand for specific products must be predicted and matched to production and the ability to deliver so that the two sides are in sync. There are many reasons why this may not happen: products may not be produced or delivered in time or may not be delivered in the required mix (for example of colors, sizes, quantities and so forth), may be misplaced, may be stolen by employees or externals or may have expired. Making an inaccurate prediction in excess of true needs can also have negative effects since the extra stock will not be sold and will have to be discarded at a loss. Finally, there are performance issues that are inherent to the modus operandi of the supply chain itself, primarily related to the time lag between ordering and delivery. For example, in cases when demand fluctuates considerably and cannot be met responsively, it is common practice that products are ordered in excess of what is required so as to maintain a stock buffer. Unfortunately, such safety stock orders create false demands lower in the supply chain, which are amplified downstream and result in wasted effort and resources – this condition is often referred to as the bullwhip effect.

To provide good performance, it is necessary that SCM addresses the following tasks:

- Distribution network configuration, that is, how to structure all levels of the supply chain network including the selection of suppliers, the number and location of production facilities, distribution centres, warehouses and retail outlets.

- Distribution strategy, that is, the organization of transportation of products between the different links of the distribution network. Options available to SCM are centralized versus decentralized coordination, direct shipments, cross docking between trading partners, pull or push strategies and the use of third-party logistics.

- Inventory management, that is, how to ensure that records of the quantity and location of inventory levels are accurate and updated in a timely manner, including raw materials, work-in-process and finished goods.

To be sure, to effectively conduct these tasks, SCM requires detailed information management and coordination across business boundaries throughout the supply chain. As a result, it is a particularly critical component in implementing any SCM strategy as this scale is the effective use of information technology. Despite the fact that to a certain extent SCM is about processes, training and business partnerships, it is inconceivable that its objectives can be achieved to any significant extent without computing and communications. In particular, it is necessary to integrate systems and processes, taking into consideration the complete structure of a particular supply chain, to share information including demand signals, forecasts, inventory and transportation and reduce delays in transmitting this information between trading partners.

3. Business Computing and the Supply Chain

Thursday 29 November 1951 marks the beginning of business computing. Before that, computers had only been used in scientific and military applications. On that day, at the offices of J. Lyons & Co[1] LEO, the Lyons Electronic Office, became the first ever software used to conduct business. LEO was able to calculate the amount and cost of raw materials required to meet the nationwide orders for bread placed with the company [10] and initiated a trend for computers to support and improve the efficiency of business processes through a detailed understanding of the objectives of business users.

Supply chains offer great variety ranging from supplying fresh food from the farm to the supermarket shelf, to delivering uniforms from the manufacturer to the soldier in the desert. Yet, they all share the same objective: to keep the process simple, standard,

---

[1] J. Lyons & Co. was founded in 1887 and grew to become one of the largest catering and food manufacturing companies in the world. At its peak, Lyons owned the popular Baskin Robbins and Dunkin Donuts brands, but in the 1970s the company was severely affected by high interest rates and finally became defunct in 1998.

speedy and certain [32]. To achieve this goal, it is necessary that all trading partners across a particular supply chain exchange information frequently and accurately, that supply chain costs be minimized, and that all goods and services moving through the supply chain be unequivocally identifiable at all times. An essential element to any solution that can meet these requirements is the use of open, worldwide data standards for globally unique product identifiers and a universal product classification system, combined with internetworked information services that can be used to track and trace goods and services.

Automation in open supply chains is becoming even more important due to the increasing use of RFID which can provide the required high product visibility and the free flow of information into fully automatic systems that can identify product items and link them to their associated information without any manual input. This level of interoperability through direct machine-to-machine interactions at such large scale demands the availability of open shared specifications describing every aspect of business activity.

In the decades since LEO became operational, two ingredients in particular have played a central role in facilitating such automation: the availability of standard product identification and classification schemes and the ability to exchange messages about business processes between trading partners across a supply chain in standardized formats. Unique product identification in particular has become ubiquitous and highly visible through the popularity of bar codes which are exactly representations of such identifiers. Moreover, the majority of transactions between trading partners is carried out through some dialect of the Electronic Data Interchange (EDI) standard, which defines templates for common business actions, for example ordering and invoicing. We discuss each of these developments in turn in the following sections.

3.1   Unique Product Identification

Tracing of the history of business computing in the supply chain identifies a second landmark date as Wednesday 26 June 1974, when the first bar code was scanned and the collected identifier used for a commercial transaction. This was the culmination of a long process that lasted over 30 years to develop automated ways of capturing product data. Since then, supply chain automation has grown rapidly and the use of bar codes has spread from retailers to suppliers and ultimately to the suppliers' supplier.

The history of modern bar coding began in the 1940s, when in response to a challenge by the president of an American food chain, Woodland and Silver of Drexel University, created a system to encode information in combinations of concentric circles printed on paper. At that time, their solution was limited by the inability to automatically input the encoded product identifier in a computer system. This problem was not addressed until the mid 1960s and until the advent of lasers which made reading bar codes practical. The initial idea received little attention in the grocery sector until 1968 when RCA, which had acquired the intellectual property, developed a similar symbol and corresponding scanner and tested it extensively during the early 1970s [3].

Bar coding was also investigated in the rail industry as a means of tracking individual railway wagons. By 1962, Sylvania Corporation introduced a system using optical scanning devices to read orange and blue coloured bars on a non-reflective black background. By 1968 the colours were eliminated, and by 1971 about 95% of all railway wagons had been bar coded. At that point, only 120 scanners had been installed, and recession in the mid 1970s led to the system being abandoned.

Owing to such diverging activities, it soon became apparent that separate groups would develop different and incompatible systems for product identification that could considerably hinder the wider acceptance of a common standard. As a result, in 1969 the American National Association of Food Chains (NAFC) proposed a product-marking system to representatives of all sections of the grocery industry, including manufacturers, retailer, and retail associations. The result of these efforts was the recommendation in 1973 by the Ad Hoc Committee of the Grocery Industry of the Universal Product Code (UPC), a common standard for the representation of the information held in bar codes. By the end of 1973, over 800 manufacturers were assigned UPC numbers, and the following year scanners from IBM and NCR were supplied to retailers. It was such a UPC code that was used in 1973 for the first bar-code-based transaction.

3.2   Universal Product Identification

The original UPC was a ten-digit code, with five digits used to identify the manufacturer and another five for the product line, and also a symbol design that would be printed on products was defined. A core management activity under the scheme is the allocation of prefix numbers to companies, to manage the numbering space and ensure that each number is unique. This task was assigned to the Uniform Grocery Product Code Council established for this purpose in 1971 and became Uniform Product Code Council in 1974 by which time it had over three thousand members. Since 1984 the Council is known by its current name, the Uniform Code Council (UCC).

Naming this solution, the *Universal* Product Code was of course an exaggeration. Not only was it not universal, but it did not even extend beyond North America. Soon after their introduction, these ideas were taken over by European retailers and manufacturers, who made it truly international. Moreover, they were extended and developed in several ways, for example, where UPC concentrated on the point of sale, the European approach adopted a supply chain perspective and code semantics were further developed beyond the manufacturer/product identification pair.

This work was carried out by a core group of collaborating companies, which formed was for this reason in 1977 under the so-called European Article Numbering (EAN) system. EAN worked closely with its national counterparts such as the UK-based Article Number Association (ANA). Such collaboration was uncommon within the fiercely competitive consumer goods sector and was the result of the clear need to adopt common open standards.

One of the new features of the EAN system that make it particularly flexible is the separation of data from the data carrier, that is, the product identifier from its bar code representation. This feature has enabled the introduction of more types of bar code symbols in addition to the original EAN specifications. For example, RFID tags can be used to encode existing EAN product numbers and this is indeed the method of choice for the ISO item-level tagging standards which we discuss in the following section. In any case, the focus on *item identity* rather than *product information* in automatic data capture has provided great adaptability and efficiency over the years, which seems to suit well current technologies.

EAN extends well beyond Europe and to mark this orientation in 1981 EANA was renamed as International Article Numbering Association (IANA). EAN codes are the standard product identification scheme across the world except North America, where UPC is still the dominant form. Several provisions ensure that the two systems are compatible, notably the formal agreement in 1990 between EAN and UCC to co-managed global standards for identification of products, shipping units, assets, locations and services, as well as a variety of other business standards that have become known as the EAN.UCC system. To complete the integration of UCC within EAN International, the organization was re-launched across the globe as GS1.

3.3 Anatomy of a Bar Code

Looking closer at a typical bar code for example, the one following the EAN-13 standard[2] displayed in Fig. 2, it is a symbol which encodes strings of 13 decimal digits, which represent unique identifiers for specific products following the Global Trade Item Number (GTIN-13) specification. This symbol can be read into a computer system using a (portable or fixed) low-power laser scanner, which can translate the sequence of white and black bars into the corresponding digits.

---

[2] Other EAN schemes follow a similar structure but support different identifier lengths.

Fig. 2. A typical example of an EAN-13 bar code.

The encoded number follows a scheme designed to ensure that each number assigned to a product line is unique and includes a unique number which identifies a particular user (most commonly its manufacturer):

- The first two digits are called the indicator digits and specify the particular numbering system used. In the case of the EAN-13 bar code of Fig. 2, the indicator digits correspond to the GTIN-13 system.
- The following five digits is the GS1 company prefix, which represents the manufacturer of the product.
- The following five digits represent the product code, which identifies a product line (but not individual items).
- Finally, the last digit is a checksum used by acquiring computer systems to confirm that the code has been retrieved correctly.

The company prefix which is also known as the manufacturer code is assigned to the particular business by GS1, while the digits corresponding to the product code are selected by the manufacturer.

The GTIN number itself does not contain classification information in it – information about the industrial sector, the country or the region where the product was manufactured or the type of product (for example clothing, food, electronic device and so forth) cannot be retrieved from the code. It is a simple unique identifier akin to a key in database parlance, and to obtain associated product information, it is necessary to query a related product information repository. Moreover, the unique identifier characterizes the product, for example, one carton of 1-liter orange juice made by the Squeezed Juice company, rather than a particular instance of the product for example, the specific carton of Squeezed Juice orange juice which was produced at 12:15:01 on January 1st 2007 at the Orange Grove facility.

Note that these are many bar code varieties, of which several of these outside the EAN.UCC system and some of which carry additional information, for example, sell by dates or product weight, or designed to deal with specific environments including pallets, locations and returnable assets. Specialist formats have also been employed in specific situations, for example, the datamatrix standard for small items used to mark surgical instruments, and new higher capacity symbologies have also been introduced, some of which employ colour and three-dimensional structures.

3.4   Electronic Data Interchange

The second core ingredient of modern supply chain management information systems unfortunately is not associated with a specific landmark, but has come about as a process rather than as a single event. Electronic Data Interchange (EDI) is the ability of direct computer-to-computer transactions between vendor and ordering systems, for example, to place orders, create invoices and reconcile transactions. EDI has very considerable advantages over paper-based procurement systems since it can reduce the time needed for product replenishment, labour costs, accuracy and access to information. The final point is of particular importance: by recording detailed information about patterns of consumption over time, it becomes possible to develop an accurate model of product use and a strategy for product movement through the supply chain.

Development of EDI started in the early 1960s as a response to the perceived need for a common vocabulary of business exchanges. Of particular relevance to the current discussion is the work carried out under the remit of the United Nations Directories for Electronic Data Interchange for Administration, Commerce and Transport (UN/EDIFACT). Unfortunately, the resulting system has been particularly complex and overloaded, hard to deploy and often leads to unnecessarily irksome implementations. As a result, several groups have identified and promoted the independent use of particular subsets that satisfy the needs of specific industrial sectors, specific business processes or specific supply chains. For example, GS1 has developed EANCOM to support cross-border trade and cover only the functions required to effect a complete trade transaction.

Another case of a partial EDI vocabulary within a specific market segment defined with the EAN.UCC system is the Trading Data Communications standard (TRADACOMS). TRADACOMS was developed in the early 1980s and employs EAN codes for product identification. Similar to other EDI activities, TRADACOMS came about as a response to the desire of several leading retailers in the UK at the time to establish electronic communications with their suppliers, which was failing due to different and incompatible message structures and content used by each company. Successful implementation of TRADACOMS in trials allowed electronic invoicing to become supported in law, and indeed the system is still widely used in retail applications.

3.5 The GS1 System

The benefits of the common product identification schemes and business message exchange formats outlined in the previous sections highlighted the advantages of an open and standard supply chain management system, but fall short of providing a complete solution. The incorporation of GS1 as a global umbrella organization for such activities provided the structure for the formalization of the so-called GS1 System (One Global System), which aims to support the efficient operation and management of supply chains and in this way create added value for the consumer. This objective is addressed through the provision of the technological foundation for the construction of inter-operable systems for asset tracking, traceability, collaborative planning, order management and logistics across all the organizations participating in the supply chain. GS1 standards address three areas:

- Part I deals with unique identifiers for products, companies and so forth and data standards for attribute encoding.
- Part II relates to the encoding of this information into data carriers such as bar codes and RFID tags.
- Part III sets data standards for automatic electronic communication through supply chains, including conventional EDI standards (mostly employed in closed networks) as well as the ebXML family of standards for open supply chains.

ebXML in particular is a recent development which employs modern technologies including the Unified Modelling Language (UML) and the Extensible Markup Language (XML).

In practice, GS1 is a complex system in perpetual development, which affects a large business community coordinated by more than 100 national organizations operating across 133 countries. Over a million member companies worldwide use GS1, and every day more than five billion transactions are made using GS1 standards. GS1 national organizations play a critical role within this community: they help members implement current bar coding systems and business-to-business communications such as EDI, and they also represent their corresponding countries in international initiatives for new standards and solutions. Notable recent additions to the GS1 standards include reduced space symbology (RSS) bar codes, radio frequency identification (RFID) tags and the EPCglobal network.

3.5.1 Messaging for Open Supply Chains

The design of EDI is limited by its focus on closed, proprietary networks and as a result in many ways it is not suitable for use over the Internet. This is primarily due to the fact that it was designed primarily as a one-to-one technology and lacks flexibility. Moreover, the requirements for the development and operation of an EDI-based system have proven in practice to be quite significant and hardly affordable by small- and medium-sized companies, which until recently have been largely excluded from participating in electronic data exchanges as a result.

To address these restrictions and to capitalize on the business opportunities opened up by the Internet, ebXML has been introduced as an altogether new messaging technology for GS1 under the Organization for the Advancement of Structured Information Standards (OASIS). Unlike EDI, ebXML assumes that the communications substrate is the Internet and aims to provide a modular rather than a rigid set of specifications for conducting business. The use of open and well-understood Internet standards implies that ebXML can be implemented at relatively low cost due to the fact that it is supported on commodity internet platforms.

Nevertheless, ebXML is a very extensive set of specifications with universal scope both in terms of geography and industrial sector [27] and is structured around the following parts:

- **Messages:** ebXML messaging functions directly extend EDI functionality and follow the standard Simple Object Access Protocol (SOAP) envelope-and-message format.

- **Business Processes:** ebXML offers standard models that capture the flow of business data among trading partners recorded using UML. This systematic definition of specific business processes is then used as the basis for common message sequences across industry boundaries. Several such processes have been recorded in detail.

- **Trading Partner Profiles and Agreements:** Complementing models of specific processes, ebXML also provides systematic representations of company capabilities to conduct e-business in the so-called Collaboration Protocol Profile (CPP). Using the CPP, a company can list the industries, business processes, messages and data-exchange technologies that it supports. Trading partners use such CPPs to specify Collaborative Protocol Agreements (CPA) that define the business processes, messages and technologies employed.

- **Registries:** Registries are ebXML-shared repositories that hold descriptions of industry processes, messages and vocabularies used to define the transactions exchanged with trading partners in CPP and CPA formats. Such repositories can be queried by other business to retrieve details of e-business capabilities for inspection so as to locate companies with the capabilities desired in forming partnerships.

- **Core Components:** Core Components (CC) are standardized XML schemas that represent the core entities involved in ebXML scenarios. CCs are lower level descriptions of the main entities that participate in business transactions and can

be viewed as the extension of more traditional GS1 data structures updated for
use by open supply chains operating over the Internet.

3.5.2   Global Product Information Repositories

The final ingredient for effective data dissemination in the supply chain according
to GS1 vision is the Global Data Synchronization Network (GDSN) specification [5].
GDSN maintains master data alignment, or else authoritative information about any
entity that can be assigned a unique identity within the EAN.UCC system including
products, prices, promotions and locations. GDSN is a database-based mechanism
(called GS1 Data Pools in GDSN parlance) of global reach that guarantees accurate
and synchronized information across supply chains.

GDSN acts as a shared electronic directory between supply chain partners used
to increase the quality of information across all supply chain activities and thus the
efficiency of transactions. GDSN is a highly controlled environment supported by
a small number of providers authorized by GS1, which are responsible for ensur-
ing that the service is available and provides good-quality information at all times.
GS1 operates the root of the directory called the GS1 Global Registry, which holds
information about the location of all participating data pools. Individual suppliers and
retailers gain access to GDSN via subscriptions to local data pools (often provided
by GS1 national organizations) and either publish or retrieve information pertaining
to specific supply chain tasks.

Product information maintained within GDSN must be organized into categories
so as to be useful and easy to access. Such structure is provided by yet another
GS1 standard, the Global Product Classification (GPC), which defines exactly such
a hierarchical scheme. At the top of this hierarchy is the Segment which represents a
particular industrial sector, for example, food, beverages and tobacco. Within a particular segment, there are one or more families which represent broad sub-divisions;
in the same example, a particular family would be milk, butter, cream, yoghurts,
cheese, eggs and substitutes. The next level in the GPC hierarchy is the Class which
represents a collection of like-product categories, for example, milk and substitutes
and at the bottom is the brick, which represents product lines. Each brick is associated
with attributes that define the specifics of the product line.

Products manufactured by a particular company would correspond to one brick that
can also be assigned GTIN numbers which, as noted earlier, are printed on bar codes
and affixed on products. The mapping between the GTIN and the corresponding brick
as well as detailed associated information about the product would be published by
the manufacturer and via its local Data Pool into the Global Repository of GDSN. So,
when a vendor receives a shipment of such items, they need only query the GDSN to
retrieve complete information about the product. This information is guaranteed to

be fully up to date and authoritative and the whole process can be completed without any manual intervention and without the need for any direct bilateral communication.

Although clearly this is a much more complex system, this approach removes all limitations inherent in closed systems like EDI and does provide a scalable information infrastructure, which is dynamic and open to all partners. The main benefit of this approach is that by federating responsibility for the maintenance of such data, it is possible to improve accuracy of orders, invoices and other business documents, to reduce the number of delivery errors, and last but not least, to reduce the administrative requirements related to maintenance of product and location information.

4. Supply Chain Optimization

Improving the performance of a supply chain depends on identification of the inefficiencies and resolution of their causes. Typically, this requires detailed measurements of performance and then implementing changes in those areas that appear to block products, services or information. Information technology can play a central role in two ways: providing the information needed to identify the causes of inefficiencies and in improving communication between partner organizations.

4.1 Causes of Supply Chain Inefficiencies

Recent research in supply chain efficiencies has identified several common problems and quantified their effects on performance, concentrating on five areas: out-of-stock, shrinkage, invoice accuracy, unsealable products and inventory accuracy [41]. Upstream supply chain inefficiencies affect the relationships of all trading partners and result in high out-of-stock conditions at the point of sale, high rate of returns and prolonged lead times. Inefficiencies in the downstream direction negatively affect demand forecast accuracy, which results in low on-shelf availability and thus loss of revenue despite the fact that products are available on site.

Preventing out-of-stock situations. Recent investigations of out-of-stocks [19] estimate their level for the retail industry to 8.3% (varying between 7.9% in the US and 8.6% in Europe). According to this study, in 47% of the cases, this was a result of erroneous forecasting and ordering; in 28% by various upstream activities; and in 25% by inadequate shelf restocking. The latter requires particular reference as in this case the required product was available in the backroom of the retail store but was not available on the shelf. Another study [17] specific to the grocery sector found that for promotional items, the out-of-stock level was almost twice as high.

Preventing Shrinkage. Inventory shrinkage or simply shrink refers to the loss of products and can happen anywhere between their manufacture and the point of sale.

In recent years, shrinkage has been identified as a serious problem [22], which may be as high as 1.7% of sales. Almost half of it is due to employee theft, but shoplifting and administrative errors also play a significant role.

Improving invoice accuracy. Inaccurate invoices have a particularly painful effect as they lead to reduction of the expected revenue and give a misleading view of the financial standing of the organization. Yet, they are not uncommon and on average they lead to deductions estimated to between 4.9 and 9.9% of annual invoiced sales [18]. Even top-10 retailers face invoice deductions averaging 5.9%. The main causes for such deductions are erroneous pricing, coupons and penalties.

Reducing unsaleables. Products may become unsaleable for a variety of reasons, with damage being the most common, followed by expired and discontinued items. Loses due to this cause amount to about 1% of sales [30].

Improving inventory accuracy. In a recent case study of inventory accuracy, over 70% of SKU records per store were found to be in error [35]. These figures are based on actual inventory counts at six stores in the US (each representing in excess of 9000 SKUs) conducted specifically for this study and compared against the records held. Higher than actual quantities were recorded for 42% SKUs and lower than actual quantities for 29%. For an average inventory of 150 000 product items per store, the total difference was 61,000 items or else about 7 items per SKU.

Among all retail sectors, supermarkets are the most competitive as they operate with minimal profit margins. It is then even more important for grocery retailers to exploit any opportunities to reduce the inefficiencies outlined above wherever possible using information technology. Over the past fifty years, they have certainly pursued this objective with considerable success.

4.2 Efficient Consumer Response

Grocery products and/or Fast Moving Consumer Goods (FMCG) have been one of the main beneficiaries of the improved understanding of the structure and performance of supply chains. The development of strategies that employ this new understanding to achieve improved performance has become possible through the use of technology, notably bar codes, messaging and resource planning and optimization software. Implementation of these techniques in the field requires extensive coordination between trading partners and to a large extent is orchestrated by Efficient Consumer Response (ECR), a voluntary industrial initiative to raise performance levels across the entire retail sector [31]. ECR promotes the premise that improvements will be made as a result of the continuous and detailed self-examination of processes and procedures across the sector, the development of concrete guidelines and recommendations and by closely promoting their implementation. ECR was initiated in the United States but its perceived advantages from a business perspective have extended its scope to the rest of the world, with national and regional initiatives in action.

ECR has developed a specific strategy around three objectives:

- to increase consumer value,
- to remove costs that do not add consumer value, and
- to maximize value, while at the same time minimizing inefficiency throughout the supply chain.

In practice, these priorities are used to identify and fulfill specific goals, for example, providing consumers with the products and services they require, reducing inventory, eliminating paper transactions and streamlining product flow. To meet these goals, distributors and suppliers are making fundamental changes to their business processes that can only be enabled through the implementation of novel information and communication systems.

4.3   Information Flow and ECR

Nevertheless, fifteen years of ECR involvement and the introduction of information systems in production and logistics control have not completely removed inefficiencies in modern supply chains, which directly impact retail operations. Despite the fact that information is shared between trading partners more frequently and in finer detail, such exchanges are still not adequate to provide the required accuracy of demand forecasts and thus the scheduling of the replenishment process. Indeed, changes in patterns of consumer demand change frequently but propagate relatively slowly through the supply chain. As a consequence, upstream partners have an inaccurate, time-delayed view of the current situation, which is often the cause of the bullwhip effect discussed previously. Another direct consequence of low-demand forecast accuracy is that trading partners have to maintain increased inventory levels as a security measure in response to unpredictable increases in demand, which further increased warehousing and logistics costs.

In practice, it is still common to forecast consumer demand by processing historical point-of-sale data, using decision support systems that utilize data warehousing and data mining techniques. One core limitation of forecasts conducted in this way is that they are not effective in taking into account the influence of promotions and other marketing instruments since the success rate of such mechanisms is generally hard to quantify beforehand. Even when the use of real-time point-of-sale data is possible, forecasts still have lower accuracy because demand patterns are changing rapidly and such fluctuations cannot be captured in a timely manner at the point of sale but have to be identified earlier in the consumption process.

One approach developed within ECR to address the problem of accurate forecasting is the so-called Vendor Managed Inventory (VMI) where the vendor, rather than the customer, specifies delivery quantities sent through the distribution channel [40]. This reversal of roles in the procurement process has become possible through the deployment of EDI. VMI had succeeded in reducing stock-outs and inventory buffers in the supply chain. Common benefits of VMI implementations include a significant reduction in supply chain length, the centralization of forecasting and frequent communication of inventory levels. VMI has a particularly noticeable effect on fleet management since the order in which delivery vehicles are loaded is defined by the system with items that are expected to stock out have top priority, then the items that are furthest below the targeted stock levels, then advance shipments of promotional, and finally, items that are least above targeted stock levels.

In addition to EDI, VMI also depends on the common use of universal product identifiers and bar codes to record and process shipments with only limited manual intervention. Bar coding in particular is essential for the automated initiation and entry stages of the order cycle and can reduce the total cycle time by several days at a time. When used together, standardized messaging and bar codes can enable collaborative relationships in which any combination of retailer, wholesaler, broker and manufacturer can work together to seek out inefficiencies and reduce costs by looking at the net benefits for all participants in the relationship. Such techniques work at the Store Keeping Unit (SKU) or container level, for example, a case, a pallet or a truck. However, an inherent limitation of existing SG1 bar code schemes is that they cannot differentiate between two SKUs from the same product line. As a result, the specifics of a particular SKU cannot be recorded unambiguously and so large inaccuracies in inventory levels can be observed [24].

Overall, VMI has been successful in significantly reducing inventory levels and the number of stock-outs. The latter issue is particularly important not only because of lost sales but also because shelf availability is central to supermarket strategy. Indeed, a significant proportion of supermarket profit margins are due to interest-free periods for products already available on the shelves. Thus, one of the main concerns of retailers implementing VMI has been the perception that reduced inventory will result in less product being available on the shelves at any one time and therefore loss of market share. A partial solution to the problem is to fill shelf space with other SKUs from the same vendor, but this approach does not fully address the problem.

The quality of information flow between trading partners can be improved in two ways that can have significant impact:

1. By extending unique identifier schemes at the containment level and in such a way that different instances of the same type of SKU can be unequivocally

identified. In this way, individual containers can become traceable and associated with location and other related meta-data. Moreover, the concept of identification can be taken into the next level and schemes that identify uniquely specific product items can be developed, thus assigning a single identity to a particular product item.

2. By fully automating the product identification process so that the need for manual operation is removed. This can remove a variety of errors in data input, and also with the appropriate hardware provisions, it can supply faster and more points of control across the supply chain.

Both of these improvements can be achieved using RFID technology which we discuss in detail in Section 5. A more ambitious approach to improve forecasting accuracy involving RFID would aim to capture information much earlier in the consumption cycle, for example, when products are removed from the display, or even earlier when already purchased products are used by the consumer and their packaging discarded, thus initiating the replenishment process. The latter approach would require fully automated unique product item – rather than container or SKU – identification and is further explored in Sections 8 and 9.

5. RFID Technology Basics

Although RFID is a relatively simple technology, it offers a unique advantage in that it allows highly compact battery-free electronic devices, the so-called tags, to be embedded in objects, artifacts, locations or living organisms and automatically identify their carrier using wireless communication and without any need for manual processing. Generally, this identification information would be a code that would uniquely pinpoint the carrier within a numbering scheme. In some cases, a tag would also hold and transmit a small amount of additional data associated with it. The information help in a tag is retrieved by a higher capability device called the reader which transmits power to the tag and directs the communication. As a result, RFID is never used in isolation but it depends on a variety of supporting information technologies to create usable systems.

In this section, we will consider each element of a complete RFID system, but before delving into the details it is worth exploring how the different components fit together. Unlike other wireless communication systems, RFID is asymmetric in that the tag and the reader are devices with very different characteristics that take distinct roles in the process. In addition to readers and tags, an RFID system would also have a number of associated services which provide the reader with a scan plan and receive

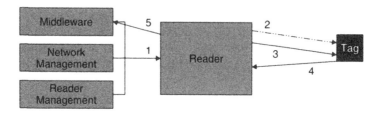

Fig. 3. Components of a complete RFID system and sequence of events.

the results of the actions specified. The sequence of operations follows the following common pattern which is depicted in Fig. 3:

1. An observation plan is programmatically specified by the system developer and implemented in purpose specific middleware, which relays the instructions to one or more readers for execution.
2. Upon receipt of the observation plan, the reader starts transmitting with the immediate effect that tags within its vicinity receive power which they can use to power up.
3. After conducting an inventory of all tags that are within range, the reader selects a specific tag according to the parameters specified in its observation plan and interrogates it specifically for its product identifier and possibly associated information.
4. The tag receives instructions, checks the contents of its memory and responds to the query of the reader (the actions in steps 3 and 4 may be repeated several times per second).
5. The responses from all relevant tags are processed, filtered and aggregated by the reader and a report is returned to the middleware or some other consuming application.

In addition to the actual RFID processing steps, the reader would often communicate with the network management software to report its status and also with reader-specific management software that would monitor operational parameters specific to RFID, for example, the correct operation of all antennas attached to the reader.

5.1   Operating Principle

Despite its numerous applications, RFID is a relatively simple technology which allows for the short-range wireless transmission of small amounts of information, often representing a single identifier that gives it its name. As noted earlier, RFID is asymmetric in that communication is established between peers with distinct

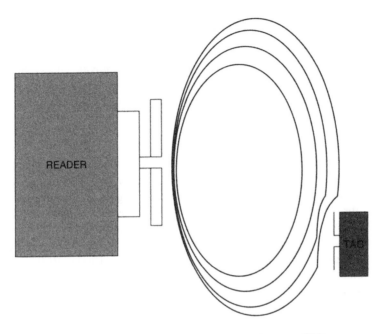

FIG. 4. Communication by reflection in ultra high-frequency RFID tags.

roles: one peer, the so-called reader or interrogator, takes on the role of the transmitter and the other, the so-called tag, the role of the responder.

This split of roles allows the communication of the tag by modulation of the electromagnetic waves emitted by the reader instead of creation of its own transmission (cf. Fig. 4). This approach implies that a complex reader can be used with a very simple tag of small size, which can be built at very low cost. Moreover, in the case of passive RFID tags, electromagnetic waves emitted by the reader carry enough energy to be used by the tag (using the coupling effect induced on the tag antenna by the electromagnetic carrier wave) as its source of power.

These two core ideas behind RFID, namely, communication by reflection and remote activation using radio frequency, were first discovered in the 40s and the 60s, respectively. But it was not until the mid 70s that fully passive relatively long-range systems became possible (for a more detailed discussion of the history of RFID, see [29]); however, early tags were still limited by the non-availability of high capacity, high-performance chips. At that time, RFID could only provide up to a dozen read-only bits on massive die sizes which occupied most of the tag volume. Shrinking electronics, especially in the 90s, have been critical to the development of the current generation of tags which are both significantly more power efficient and provide higher storage and computational capability – both as a result of miniaturization.

5.2   RFID Types

One particular type of RFID, the so-called active tag, uses batteries as their source of power and is not wholly dependent on the reader to provide energy. Such tags have considerable advantages over passive tags that draw all their power from the reader signal, as they transmit at higher power levels and thus have longer range and support more reliable communication. Moreover, active tags can operate in particularly challenging environments; for example, around water, it is easy to extend them with additional sensing capability, for example, temperature sensors, and they can initiate transmissions, but they stop operating when their battery expires. Despite their advantages, the current interest in RFID is solely due to passive tags which do not depend on batteries and thus do not require recharging or replacement. Active RFID, on the other hand, is just one of an increasing number of wireless local area communication technologies and as such it is of limited interest to this survey. In this review, we only consider passive tags as they are the only viable solution for large-scale deployments. For this reason, we will refer to passive RFID simply as RFID, without further qualification.

RFID tags can be naturally classified under two main categories: those that use the magnetic component generating the near field of the radio wave and those that use the electric component, which generates the far field (cf. Table I for a comparison of their characteristics). Near-field tags communicate by changing the load of the tag antenna in such a way that they control the modulation of the radio signal in a process appropriately called load modulation. These changes can be detected by the reader and decoded by examining changes in the potential variation in its resistance. Because the magnetic field decays very rapidly with distance from the centre of the reader antenna (inverse cube ratio), the changes to be detected by the reader are tiny compared with its own transmission. For this reason, the tag modulates the radio signal in such a way that it responds in a slightly shifted frequency from that of the reader (what is often referred to as the sub-carrier frequencies).

Power transmission from the reader to the tag is by magnetic induction (the principle employed by power converters) and for this reason near-field readers and tags have a characteristic antenna design that also makes them easily identifiable: their antenna is a simple coil. The effectiveness of this process depends on the strength of the near field at the tag location, which in turn depends on the distance between the centre of the reader and the centre of the tag antennas (and the particular frequency used). In any case, at frequency f, the near field ends at distance proportionate to $\frac{1}{2\pi f}$ from the reader antenna. For example, at 13.56 Mhz, the frequency used by the popular ISO 14443 standard, the near field extends to about 3.5 meters from the reader. However, in practice ISO 14443 systems would consistently work at a maximum range of approximately 30 cm using medium-sized antennas on the reader (radius approximately 20 cm) and credit-card sized tags.

One of the advantages of the 13.56 Mhz frequency that makes it so popular is the fact that this section of the wireless spectrum is assigned worldwide to smart cards and labels and hence it is globally available to the vast majority of RFID applications. Other frequencies commonly used by near-field RFID are within the 120–136 kHz range, but these are loosing rapidly in popularity as they can only be employed for very short-range communications. Their short range makes them unattractive for applications as in most practical situations they necessitate contact of the card and the reader (but not of the electronics directly).

RFID systems using the far field of the carrier wave operate using a technique called backscatter rather than load modulation. This process is very similar to the operation of the radar in that the tag reflects back a small part of the electromagnetic wave emitted by the reader. The reflection can be used to transmit information by examining the so-called reflection cross-section, that is, the signature of the component of the wave that has been sent back to the reader and compared with the original. In practice, data are encoded by the tag by turning on and off the load connected to its antenna and thus shifting the reflection cross-section between two clearly identifiable characteristic signatures. Similar to near-field RFID, also in this case there is very considerable loss of power during the reflection process and readers have to be sensitive to less than a microwatt in most cases.

Because of the involvement of the far field, tag and reader antennas are dipoles. This fact can again be used to identify far-field tags via simple visual inspection. Far-field RFID commonly operates in the UHF band between 865 and 956 Mhz, but the complete range is not available to applications globally (and there are also radically different signal power output limitations especially between Europe and the US). Instead, common far-field tags are able to respond in the complete range and it is the responsibility of the reader to select frequencies that are allowed within a particular regulatory region (typically 865–869 Mhz in Europe, 902–928 Mhz in the US and 950–956 MHz in Japan). Far-field systems allow for longer range communication and it is common to achieve between 3 and 4 meters using approximately 30 cm

TABLE I
COMPARISON OF HF VERSUS UHF RFID TECHNOLOGIES

	HF (Near Field)	UHF (Far Field)
Frequency	≈ 13 Mhz	≈ 900 Mhz
Spectrum allocation	Uniform	Fragmented
Cost (per tag)	< 15 cents	< 15 cents
Range	< 30 cm (1 m max)	< 4 m (10 m max)
External interference	No	Cellular phones
Memory capacity	4 Kbits	256 bits

antennas and 10 cm tags. Using larger antennas and power amplification, the range of such a system can reach up to 10 meters. More detailed descriptions of far-field RFID performance can be found in [8].

5.3 Readers

An RFID reader or interrogator consists of three main components (cf. Fig. 5):

- One or more antennas, which may be integrated or external.
- The radio interface, which is responsible for modulation, demodulation, transmission and reception. Due to the high-sensitivity requirement, RFID readers often have separate pathways to receive and transmit.
- The control system, which consists of a micro-controller and in some cases additional task and application-specific modules (for example, digital signal or cryptographic co-processors) and one or more networking interfaces. The role of the control system is to direct communication with the tag and interact with applications.

RFID readers are increasingly becoming complete network computing devices (akin to routers) that provide advance processing of RFID observation streams and wired or wireless connectivity to the internet. Such readers would receive a scanning plan from a driving application or other middleware, which they would implement by issuing state-transition instructions to the tags within their range. The latter step usually has three stages: broadcasting to all tags within range and receiving responses, selecting a particular tag as the peer for communication, and exchanging information with the selected tag. This process can be quite complex, especially in the case where

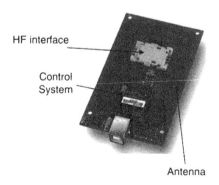

HF interface

Control
System

Antenna

FIG. 5. RFID reader and subsystems.

a large number of tags are within range or when two or more readers overlap. In such cases, additional collision-avoidance techniques must be implemented to ensure that communication is organized in a structured way so as to allow the participation of all tags in this process [11].

5.4 Tags

The tag is a far simpler device and consists of:

• The antenna.
• A capacitor that stores harvested power.
• The chip which in most cases implements a simple state machine and holds the object identifier.
• A protective paper or polymer enclosure, which guards against the rupture of the antenna that would result in the immediate expiration of the tag.

A typical example of a modern tag is the EPC Class 1 Gen 2 [8, Chapter 4] which operates at UHF frequencies (cf. Fig. 6). The chip has a relatively complex

FIG. 6. Gen2 RFID tag operating at UHF frequencies.

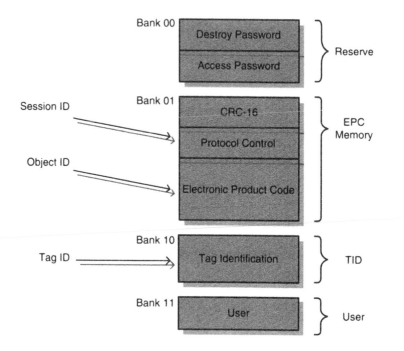

FIG. 7. Memory layout of an EPC Gen 2 tag.

non-volatile memory structure divided into four distinct areas (cf. Fig. 7). The reserved memory bank holds two 32-bit passwords, the 'access' password for gaining access to the contents of the tag, and the 'kill' password that when presented permanently disables the tag. The EPC memory bank contains the Electronic Product Code, a universally unique identifier assigned to the object, location or other asset on which the tag is attached, and optionally other metadata. The Tag Identification bank contains information about the type and the manufacturer of the tag including a unique serial number which identifies the tag itself. The user bank is optional and can be used freely by applications.

It should be clear from this discussion that a single tag holds several identifiers or codes that correspond to different functions and have distinct roles and semantics, including a fixed tag ID and a writable object ID. Tags often use a third identifier, the so-called session ID (in the case of Gen 2 tags, this is a pseudo-random number generated by the Protocol Control section), which is used by the reader to address the tag during a particular session. The session ID is roughly equivalent to the MAC address of a typical wireless networking physical layer protocol, but in the case of Gen 2 it is only locally unique. Alternatively, the session ID may be fixed and stored in the tag memory as is the case for ISO 14443 Type A tags. Note that tags that employ this approach can be easily traced using the session ID as a handler, a fact that raises very considerable privacy and security issues which we discuss in more detail in Section 10. For this reason, most recent tag protocols implement a randomization process, whereby tags use a pseudo-random number each time they are interrogated by a reader so as to avoid easy tracing.

5.5   RFID as Smart Product Labels

Although it has been noted that more than one identifiers are stored in a tag, all but one are involved in low-level operations and are thus of limited interest for enterprise computing. The object ID is the identifier that is related to the product, container or location where the tag is affixed. One clear use of this stored information is as a direct replacement of bar codes: the exact same information stored in a visual representation can be stored in a tag in electronics and transmitter over radio frequency. Even this simple substitution of bar codes with RFID provides considerable advantages, namely:

- higher capacity, so that larger identifiers or even additional metadata can be stored.
- higher data read rate, so that many more product labels can be read in very short period of time.
- the tag is rewritable, so new data can be added during the product lifetime or old data can be updated or changed to reflect changes in the product.
- greater resilience to damage, especially since the RFID tag could be embedded safely in the product fabric itself.
- greater read range and independence from line of sight requirement.
- anti-theft support, as tags can be identified at exit points.

Of course, despite these advantages, a direct replacement of bar codes with RFID tags has very considerable cost implications since bar codes are more often printed with packaging and have no cost at all.

Although valuable in some cases, such a direct substitution of bar codes for RFID would fail to capitalize on the full range of opportunities offered by the technology. Moreover, it would fail to recognize and build on top of the current generation of network infrastructures, which have advanced since the introduction of the bar code. As a result, the new circumstances combined with the capabilities of RFID offer a unique opportunity to re-think and re-design systems of unique product identification. There are several current proposals on how to best extend current schemes, and in the next section, we review some of these proposals, with particular reference to the work conducted within the EAN.UCC system.

5.6 Identifiers

The most successful numbering scheme in terms of industrial adoption so far that is specifically developed for RFID and use in the supply chain is defined within the Electronic Product Code (EPC) specifications, part of the EAN.UCC system. Unlike other generally available RFID standards, EPC defines both how and what data will be stored in the tag including the tag memory layout (as described in the previous section), for communication with readers, and for the composition and layout of a unique identifier scheme which extends existing GS1 schemes.[3] The EPC identifier in particular can follow one of several schemes, depending on whether the tag is used to identify a product container or item, a location or some other asset.

The most important type of identifier encoded in EPC is the Serialized Global Trade Identification Number (SGTIN), which comes in two version of different lengths (96 and 198 bits correspondingly). SGTIN-96 codes are made up of six parts, namely,

- Header, which identifies the tag as an SGTIN-96 (8 bits).
- Filter Value, which allows the pre-selection of the object type (3 bits).
- Partition, which indicates the split of the last 82 bits between the remaining three fields (3 bits).
- Company Prefix, which contains the GS1 company prefix (20–40 bits).
- Item Reference, which contains the GTIN reference number and identifies the product line (4–24 bits).
- Serial Number, which is the unique identifiers of the specific tagged item (38 bits).

In following with common practice within GS1, the Header, Filter, Partition and Company Prefix sections of the EPC are provided by GS1 so that their use and assignment is coordinated and guaranteed to be uniquely defined, but the Item Reference and Serial Number are assigned by the manager or else the manufacturer of the product. An example of an EPC encoding an SGTIN-96 and its interpretation is displayed in Fig. 8.

ECP also provides schemes for tagging other types of resources in addition to product items, including shipping containers (for example, pallets and other SKUs), returnable assets (for example, fruit cases) or general asset items and locations. In addition to these identifiers defined by GS1, there are also provisions for the inclusion of general-purpose identifiers within EPC as well as resource identifiers following the Department of Defense numbering schemes.

---

[3] The specification also includes a Filter Value which is not part of the identifier but provides a shortcut in that it is a quick way to identify the particular type of identifier encoded in the tag and is used for fast preselection of particular tag types.

HEX	30700048440663802E185523
Binary	0011000001110000000000000100100001
	000100000001100
	1100100000000000001011100001100001
	01010100100011
URN	urn:epc:tag:sgtin-
	96:3.0037000.06542.773346595

Filter	Company Prefix	Item Reference	Serial Number
3	0037000	06542	773346595
Shipping Unit	P & G	Bounty Paper Towels (15 pack)	Item UID

FIG. 8. Example of an EPC SGTIN-96 tag and its decoding. The top table shows the actual forms of the EPC in different stages of the encoding process and the bottom shows the interpretation of the SGTIN-96 identifier in particular.

Looking closer at the Serialized Global Location Number, this identifier is a serialized form of the Global Location Number (GLN) defined within the standard EAN.UCC system and includes provisions for an extension serial number that represents internal company locations that are not openly available to external parties. SGLNs follow a very similar structure to SGTINs with header, filter, partition and company prefix. The last past of the GLN is the location reference, which is a number, the semantics of which are at the discretion of the manager. Since these numbers cannot be interpreted without access to their definitions, it is necessary for a company to publish the appropriate correspondence in a publicly available location, which is often the GDSN.

One point that sets GLNs apart from other similar systems is that they define a rather extended concept of location in addition to physical places, which in the context of the supply chain would often be stores, warehouses, manufacturing plants, warehouse gates, loading docks or vending machines. GLN also includes within its scope legal (for example, companies, subsidiaries or divisions) and functional entities (in most cases, these would be departments within the company, for example, accounting or fulfillment). In any case, this unique identifier can be encoded in an RFID tag which can be automatically read by interrogators within its vicinity, which can subsequently resolve this information through the GDSN and thus discover their location.

The Serial Shipping Container Code follows the common structure, with the notable exception that its serial number segment is defined by the standard EAN.UCC systems. Similar structure is also followed by the final two types of indemnifiers called Global Returnable Asset Identifier (GRAI) and Global Individual Asset Identifier (GIAI). Finally, EPC provides for two additional types which are defined outside the EAN.UCC system, namely, the resource codes defined by the Department of Defense specification for military supply chains and a general-purpose type predictably called General Identifier (GID-96), which is a catchall for other uses of the EPC tag specifications.

A competitive scheme to EPC is the ISO/IEC 15459 specification on unique identifiers with provisions on registration (Part 2), common addressing rules (Part 3), transport unit address provisions (Part 1) and item-level tagging for the supply chain (Part 4).

Under this scheme, a guaranteed world-wide unique serial object identifier (i.e., the object ID) is associated with an artefact by its manufacturer at production time. ISO 15459 codes have four parts: data identifier (DI) header, issuing agency code, company ID and serialized item code (cf. Fig. 9). In conformance to previous related ISO standards, each part of the code holds alphanumeric digits rather than numbers. The DI specifies the structure of the contents of the object ID and follows the specification of ISO/IEC 15418 encoded under ANSI MH 10.8.2 provisions. For example, DI set to 25S specifies that the object ID is a globally unique serial object number, and DI set to 2L specifies that the object ID is a location specified in a format defined in a subsequent field, for example, a post code. Rules for the coordination of the address space are also defined in the standard, with the Netherlands Normalization Institute being the only authorized registrar that can assign IACs. EDIFICE, an association of electronics suppliers, is such a registered issuing agency and can thus provide its members with their individual unique company identification numbers. Each member can then decide internally on how to structure the object serial numbers. A common approach is to separate the number into two parts, the first identifying the type of the object – often referred to as product class – and the second identifying the particular item within this class – often referred to as item serial number.

An important feature of ISO 15459 is that unlike EPC, it accommodates a variety of existing product classification schemes that can be used as object identifiers. For example, the currently most popular way to tag objects is by way of a barcode, mostly

Data Identifier	Issuing Agency Code	Company	Serial Number
25S	LE:EDIFICE	E999	C204060897294374

Fig. 9. ISO/IEC 15459 worldwide unique serial identifier example. The object ID stored in user memory of the tag is 25SLH:EDIFICEE999C20406089.

using identifiers specified with in the EAN.UCC system that are excluded under EPC. This approach also allows the incorporation into the system of a number of other domain-specific numbering schemes under a unified hierarchical classification. For example, ISO 14223-2 defines a code structure specific for use for animal tracking, including information on the species and the premises where it is held. These codes are incorporated under ISO 15459 simply by setting DI to 8N. This facility also allows improved interoperability with other competing or emerging numbering schemes which can be incorporated under particular DIs as well as provide flexibility for future extensions.

Although not evident from the previous descriptions, EPC also supports interoperability with ISO standards although at a lower protocol layer. Gen 2 tags provide a parity bit as a toggle to indicate the type of identifier stored in their EPC memory bank (cf. Bank 01 in Fig. 7, the Numbering System Identifier is part of the Protocol Control section) so that other numbering systems can be used instead of EPC. Although the EPC scheme clearly has few differences with the previous ones and indeed several limitations when compared against ISO, it has nevertheless attracted considerable interest due to its exclusive supply chain focus and the fact that it provides a complete set of specifications for middleware, resolution, discovery and repository services (cf. Section 6). Moreover, several IT vendors have already integrated these specifications with their products and as a result, the EPC standards have gained considerable advantage against competitors.

6. RFID Software and Network Services

A recurring theme in the discussion of modern GS1 standards is their dependence on the Internet for disambiguating the semantics of the different types of identifiers that are retrieved either from bar codes or from RFID tags. Until now, we have only considered static data, that is, mappings between identifiers and their representations which are defined at the time of manufacture and do not change over the lifetime of the product. Such data are well served by the repository and network infrastructures developed for GDSN that can provide pointers to authoritative information.

However, GDSN is limited in one particularly important way that is critical for effective supply chains, namely, in that it does not trace products as they move from trading partner to trading partner and from location to location. Rather, the GDSN maintains general information about product lines and their attributes including pricing. The capability to do so is clearly fundamental in monitoring the flow upstream or downstream. To this end, in addition to GDSN, a complimentary set of network services are defined within the EPC specifications that target information related to specific product containers and items and their complete history as they cross the supply chain [16].

6.1 Middleware

One immediate implication of the construction of such a network is its massive size: the scope of the network is for every single product manufactured everywhere in the world to be tagged and tracked. Clearly, this process generates enormous quantities of data that must be available online for querying by all participants. As a result, it is necessary that a core feature of the network mechanism is that it reduces the volume of information that propagates between systems. One way to achieve this is by recording only events that make sense at the business level rather than, for example, every sighting of a particular tag.

Recall that communication is always initiated by RFID readers that may scan for tags several hundred times per second. As a result, a particular product may be observed by a certain reader several times although its condition has not changed. Keeping a record of all these observations would be unnecessary and would not provide any useful information. Instead, such raw observations should be aggregated and filtered into higher level events that are significant. This is the role of RFID middleware which provide exactly this functionality. Moving in sequence from the lower level where observations are acquired by a reader towards application-level processing, RFID captured data enters the following stages:

- **Collect observations:** Readers interrogate their vicinity for the presence of tags and subsequently request and retrieve object IDs and potentially additional data stored in the chip memory (some systems would require an intermediate authentication step to allow access to this information). Depending on the application, the duration of the interrogation cycle can vary considerably. For example, for e-passport applications, a read cycle could last up to a minute, while in supply chain applications, several hundreds of tags would be read per second. The read phase could be followed by a further write cycle as is the case in ticketing applications where information about the current trip would be added to the ticket. Additional sensors and actuators may be activated at this stage; for example, temperature sensors could be used to record the environmental conditions in which a particular object has been observed and LED displays could be operated to indicate the state of the object.

- **Smooth observation data:** Raw observation data can be erroneous and incomplete as a result of read errors. Smoothing observations is the process of cleaning the collected data from incomplete reads that are discarded, from IDs recorded due to transient and thus irrelevant objects that must also be removed, from indeterminate reads must be resolved (for example, using authoritative records from local persistent storage), and last but not the least tags that have not been read must be rescanned.

- **Translate observations into events:** Following smoothing, observation data are still not useful to applications which are interested in higher level events. For example, in a supply chain application, it is not relevant to the business logic layer if a tag has been read by a particular reader but rather the fact that a specific pallet containing particular product items has entered the warehouse through a specific portal. This transformation of lower level observations into higher level application events is typically achieved via filtering and aggregation.

- **ID resolution and context retrieval:** Specific object IDs recorded in observations and events must be associated with object descriptions and related contextual-use retrieved data. This conversion requires access to network services that play a two-fold role: (i) to map object IDs to network service locations that can be further queried about object details and (ii) to respond to specific queries related to the current condition, the properties and the history of the object.

- **Dispatch and processing of event data:** Application-level events must be returned to consuming applications for further processing. For example, a pallet entry event would trigger updates of inventory records to include the items contained in the identified shipment.

Of course, this process works bi-directionally, that is, applications control data flow by defining events of interest and by declaring their interest to the RFID infrastructure. An orthogonal layer to the application execution profile is infrastructure management, that is, maintaining configuration and status information related to the operating condition of RFID readers and other sensor elements [7].

The sequence of tasks outlined above is carried out by distinct network segments [6]: observations are collected at the reader level outside the IP network; observation processing and event translation at the network edge by the event manager; and application logic at the network core (or data centre) level. A layer of mediation between the network core and edge is provided by the network services and other event-consuming applications, which have the role of resolving identifiers into object descriptions and the subsequent querying for associated and context data. Put together, these distinct elements define the RFID stack depicted in Fig. 10. A notable feature of this approach is the introduction of the event manager [4], which implements the translation of observations into events by:

- Bridging the IP and RFID networks by translating RFID observations into higher level events via filtering and aggregation.

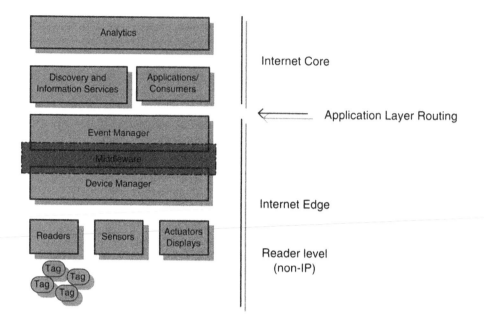

FIG. 10. The RFID stack.

- Managing the RFID reader infrastructure and related sensor and actuator devices.
- Offering a single interface to applications.

6.2 Programming RFID

Event managers require specific rules to translate observations into events. Such rules are often defined in terms of a tag scan and query plan, specified through an appropriate reader abstraction layer, which is relayed to and executed by the reader infrastructure. A scanning plan specifies the frequency of data acquisition, how many attempts are made, triggering conditions and so on. It may also include information about the specific components of each participating reader that is employed, for example, which of the attached antennas will be activated. Naturally, this device abstraction layer also provides facilities for the discovery of reader capabilities (for example, supported functionality, attached components, software versions and so forth) and can also request the pre-processing of the observation data if this functionality is supported by the reader. Finally, the device abstraction layer can also potentially support actions predicated on a triggering observation, for example, when a motion sensor detects movement. Examples of such device abstraction layers are offered by the Reader Protocol [RR] part of the EPCglobal standards and the generic interface of WinRFID [34]. Particular reader manufacturers have also developed such abstract device interfaces, but these are less useful as they can only be used with readers from specific suppliers.

The event manager provides application programming interfaces for event discovery, subscription and reporting [12, 36]. This allows client applications to find what events are available and define new ones, subscribe to those of interest and receive reports with results. Events are defined over event cycles, that is, delimited time intervals over which observations are processed. Note that although observations and events are related to read and event cycles correspondingly, the event manager decouples their respective domains and provides a clear separation of scope (cf. Fig. 11). While the adoption of cycles as the main modus operandi for the event manager may appear limiting, this is not so, as in addition to defining cycles either periodically or within fixed time slots, it is also possible to have arbitrary bounds defined on triggers fired by specific observations or by software interrupts or by external notifications.

Filtering and aggregation processing by the event manager aims to identify specific patterns in the event data and to summarize data collected from different readers over several event cycles correspondingly [42]. Filters work by applying include or exclude regular patterns, that is, by setting rules that define ID lists or ranges to be included (or excluded) in the processing of observations. For example, following the EPC filtering specification, the exclusion filter epc:gid-96:18.[321–326].* encountered while processing EPC tags specifies that the product range that corresponds to product codes between 321 and 326 will not be processed, irrespective of the serial number of the objects recorded. Similarly, the aggregation pattern epc:gid-96:*.*.X.* results in grouping observations by product code and reports only the total number of observations for each class of product. Due to relatively frequent read errors,

FIG. 11. The RFID event manager.

such filtering and aggregation techniques are rather complex to implement in practice and recent work highlights the significance of statistical techniques to improve data fidelity [23,43].

The programming interface provided by the event manager can be implemented using different methods: the Application Level Events (ALE) specification [2] is a middleware specification and the Java RFID System provides the same abstraction as a language-specific implementation of a component model built on top of the Jini event management framework. While there seems to be some consensus about the desired functionality of the application event interfaces, the actual implementation of the event manager can be done in several alternative ways. These alternatives are not mutually exclusive but adapt to their operational context and explore different trade-offs between levels of functionality and performance guarantees [21, 25]. In practice, the event manager may consist of one or more distinct physical devices and logical service end-points, with the responsibility for specific tasks shared between them.

6.3   RFID Network Services

To provide full functionality, the upper three layers of the RFID stack of Fig. 10 require access to discovery and repository management services accessible on the internet. Discovery services resolve captured object identifiers into network service locations where repository services reside. Repository services in turn can be further queried via standard service profiles to obtain trace and other meta-data related to a particular ID.

Discovery services. Mapping EPCs to network service locations is a relatively straightforward task, which can be easily accommodated within current internet infrastructures. One way to accomplish this is by simply using the directory capabilities of the Domain Name System, which can support an extended collection of record types. This approach is advocated by the Object Naming System (ONS) specification within the EPCglobal family of standards, which employs the Naming Authority Record [33] to provide associations of EPC codes to Universal Resource Descriptors. Under ONS, the serial item segment of an EPC code is removed and the remainder segments reversed and appended to a pre-determined well-known domain name (as of this writing onsepc.com). Of course, one problem with this approach is that ONS inherits and perpetuates the well-known limitations and vulnerabilities of DNS, though some of these issues are addressed by the use of a single domain where delegation and updating can be handled with greater effectiveness.

ONS is limited since it only retains the most recent service location related to a particular EPC, for example, the URI published by the current owner of an artefact. This is hardly enough in many cases: in addition to the description of the current situation of the object, many pervasive computing applications need to gain access to historical use data collected during its lifetime or at least over a considerable length of time. This is not only due to the importance of context history for system adaptation but also because of a practical consideration: Object IDs are assigned at production time from the address space controlled by their manufacturer, while the artifact itself changes ownership several times during its lifetime. As a result, such naive resolution of the EPC would point to the initial owner of the identifier rather than the current custodian of the artifact and hence authoritative up-to-date information would no longer be available at the returned service location. Moreover, the full object history is fragmented over different service locations corresponding to the different custodians that possessed the artifact at different times and a single service location could not represent the complete data set.

Hence, rather than mapping an EPC to the service point provided by its manufacturer, the resolution process could alternatively point to a secondary discovery service instead, which maintains the record of the complete sequence of successive custodians, from production to the present day. This approach is implemented in the so-called EPC Discovery Service which can be registered with the ONS and provide the list of URIs of all custodians for a particular object ID. This solution to maintaining a complete trace is preferable over the alternative, whereby the current custodian would be identified via sequence of links through past holders. Such chaining is vulnerable to broken links that can easily occur, for example, if any one of the custodians ceases to exist. One broken link would be enough to result in the complete loss of the ability to trace the object history.

Repository services. The second element of RFID network services aims to manage and maintain object usage information and is provided by custodians. Conceptually, it is little more than a federated distributed database, and provisions for this task are offered by the EPC Information Service. From a usage perspective, both standards are little more than a set of web service specifications to access object-specific data repositories. Both provide methods to record, retrieve and modify event information for specific EPCs. What does stand out, however, is the massive size and complexity of such a data repository which – if successfully implemented – would be unique. This task is complicated by the complex network of trust domains, roles and identities, which requires the careful management of relationships between authorization domains and conformance to diverse access policies and regulations. Yet, these challenges are inadequately understood at the moment as neither system has attracted significant support.

One feature of such repository services that merits further discussion today is the so-called containment profiles. This technique is necessary to form single objects out of individual components and to be able to reference them directly. Consider the case of an automobile for example: it is made up of thousands of individual components, mostly sourced from third party manufacturers, which at a certain point in time come together to be assembled in a single entity. Over the lifetime of a particular car, these components will change as a result of maintenance, upgrades or changing use. In most cases, the only requirement would be that the car as a whole is identified but in others it would be necessary to identify individual components as well. The containment profile has been introduced to address exactly such time-dependent processes, and is used within the EPC Information Service to group together components that are assembled into a new entity with its own unique EPC code. The composite object has an associated creation and expiry date and its elements can be modified via related containment interfaces.

7. Practical RFID in the Supply Chain

In previous sections we have discussed at length the information requirements of efficient and effective supply chains, and how network RFID technologies can be used to provide up-to-date and detailed information about product items, containers, service locations and other assets used in support of operations. In this section, we turn out attention on how the latter can be used to satisfy the former by highlighting in practical terms how and where RFID shall be used. Note that there are significant differences when tagging is done at the container and at the item level and these will be identified and discussed.

Let us consider a typical although somewhat simplified supply chain scenario: a variety of consumer products are manufactured at a specific facility, packaged in cases, loaded an pallets and then on trucks for delivery to a retail distribution centre (DC). Upon arrival at the DC, the pallets are dismantled and individual product cases separated and stored. At a later time, product cases are picked (and in some cases loaded on new pallets) and shipped to a local retail store. At the store, products are stored in the back room and used for restocking the shelves of the storefront in response to sales. Customers pick up products from the shelf, place them in the shopping cart and take them to check out where they complete their purchase. Clearly, this is a rather long process which is carried over a potentially very extensive geographic area and involves many individuals and organizations. It is thus not practical to monitor the progress of a particular product at every point, but rather it is necessary to identify control points, where the product changes state, and employ them to update the information held.

Revisiting this scenario from an RFID perspective, at the manufacturing facility, products are fixed with individual tags encoding their EPC code including their GTIN, which contains their item-specific serial number. Individual product items are then packaged in cases which are also tagged individually using EPC and assigned with their particular SSCC (or in some cases, a GTIN representing a case of product items). At this stage, each case SSCC is associated with the GTINs of all the items it contains and this information is published on the local EPC IS. Cases are then loaded on pallets and often enclosed within some protective material, usually either cardboard wrap or transparent stretch film, and again tagged with their corresponding SSCC. A particular pallet may contain cases from different product lines which are mixed due to the specific quantities included in the order placed by the retailer. The SSCC of each pallet is also associated with the SSCCs of the cases it contains and this information is also published on the local EPC IS.

When all the necessary pallets are prepared for shipping, they are placed in a container and loaded on a truck for delivery to one or several DCs. This point offers the first opportunity to establish a control point for the movement of products downstream in the supply chain: readers located at the exit gates of the loading bay of the manufacturer facility scan the shipment as it is being loaded on the trucks and record every product item, case and pallet identifier, grouping them together and associating them with the corresponding retailer order details and DC destination. This information can be transmitted to the retailer to anticipate the arrival of the shipment.

On their arrival at the DC, pallets are individually unloaded and moved into the warehouse via a portal which records the arrival of the scanned EPC codes and cross references the recorded numbers against those expected. If the products are confirmed to be the ones expected for delivery, the warehouse management systems are automatically updated and the pallets are forwarded for storage on the facility. The same process is followed in loading the product cases for delivery to retail shops. At the shop, cases are again received, automatically checked against the expected deliveries and if confirmed, the store WMS is automatically updated.

In this process, a central role is reserved for the loading bay doors into the warehouse (cf. Figure 12), as in most cases they represent the best location to place a control point for checking and updating the flow of products. As a result, dock doors are often turned into RFID-enabled portals where pallets are scanned and where the actual items delivered can be cross-referenced and inventories updated. This location works equally well as a control point for manufacturing plants as for distribution centres and retailer stores, and for both incoming and outgoing shipments.

Looking closer at the sequence of events involved in the operation of one such portal, the retail store receiving dock, the process starts with the receipt of an Advanced Shipment Notice (ASN). This is a common EDI message which is prepared and transmitted by the DC at the time when the pallets for a particular shipment have been loaded on a truck and have left the DC warehouse. The ASN is a notification of pending delivering and is send to all parties responsible for the movement of freight from DC to store and the contents and configuration of a shipment. In this case, the ASN would contain at least the SSCCs of every pallet and possibly also of the cases

FIG. 12. Typical RFID-enabled warehouse loading bay portal.

and the GTIN of the products included (the latter as a means of providing redundancy to EPC IS). The ASN would also record the total number of pallets, address and related details of the DC and retailer and can also contain numerous other related details.

At the store receiving dock, the external motion sensor is tripped by the movement of the first pallet passing through the portal (cf. item 3 in Fig. 13). Tripping the sensor results in the publication of a sensor-event message to the ALE engine operating on the warehouse event manager. This event marks the beginning of an event cycle which instructs the readers attached to the portal (item 2 in Fig. 13) to begin collecting observations. The readers keep scanning and discover all tags marking individual products, cases and pallets. Each tag is typically discovered and read several hundred times and the observations are passed to the ALE engine either residing on the reader itself or at the event manager (depending on the model and the capability of the reader). Observations are processed according to the event cycle specification and reported to the WMS. An event cycle may be time constrained or terminated in response to a motion-sensing event tripped by the second, internal sensor (item 1 in Fig. 13).

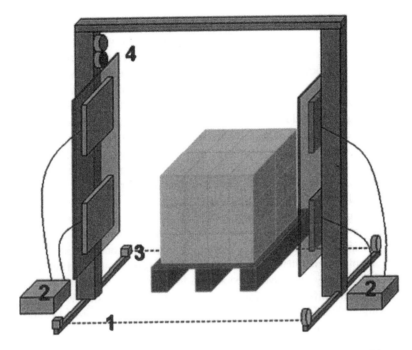

FIG. 13. Schematic of the components of an RFID-enabled warehouse portal: items 1 and 3 are motion sensor which activate on entry and deactivate on exit the postal operation; 2 are RFID readers, each of which has two external antennas attached; and 4 are red and green indicator lights that signal shipment approval or rejection.

Upon receipt of the event cycle report by the WMS, the list of products recorded is compared against the expected deliveries as specified in active ASN messages within the system. If the details match, then the pallet is expected and the portal switches on the green light on its frame (position 4 in Fig. 13), indicating that the delivery has been accepted. At the same time, the inventory is updated with the new item received and cross-checked against the relevant purchase order. In case the codes retrieved by the pallet are unexpected, the red light is switched on instead and the pallet returned to the truck.

During the aggregation cycle, the event manager filters duplicates, removes transients and codes that are not requested by the event cycle specification and returns the gathered EPC codes in a report. For example, if the event cycle specification requires that only pallet codes are collected, then all other types of tags (for example, item GTINs and case SSCCs) are observed but ignored.

Making the assumption that each product item is individually tagged with its own EPC, information gathering does not need to stop at the time when products are moved to the storefront for display and purchase. Indeed, it is perfectly feasible that a variety of locations within the storefront will be equipped with readers which will support a number of consumer applications and product demand data. For example, RFID readers at the point of sale (POS) would allow the rapid scanning of products selected by a consumer and thus a much quicker checkout which would minimize queuing time. Another related application would see readers installed in shopping carts together with embedded displays which can support a variety of personalized shopping applications, for example, e-recommendations on the basis of the content of the cart and the user profile or tracking its total cost. Last but not the least, RFID readers can be installed below shelves to monitor the number of items stored and would possibly be combined with price and quantity displays that would change automatically depending on the conditions of the product.

The latter application, for example, would have dual use, both as an assistive technology for consumers and as an effective means to monitor the availability of product on display and provide early warning of impending out-of-stock conditions. Even the simpler case where portable or hand-held readers are used to take stock at the end of the day can have considerable benefits: in case where a product is available in several different versions that are not necessarily easily discernible without checking their GTINs, quick stock taking using RFID can provide significant improvement for the replenishment process.

This is especially relevant in the case of apparel retail; for example, consider the case of a retailer of formal menswear. Suits, in particular, come in various sizes and colours, which are often identified by a single GTIN. Nevertheless, it important that a full mix of the different types is always available to fulfill consumer demand, and it is often the case that after closing, sales personnel have to manually conduct an availability survey, which is a time-consuming task that becomes particularly onerous due to the timing constraints. Instead, a quick scan of the racks you immediately identify current stock and missing ranges and colours would become automatically identifiable without further need for manual intervention. Of course, such applications would be feasible for higher cost items like garments that provide a higher return and profit margin.

8. Business Drivers

Having developed an understanding of the information requirements for effective supply chain optimization in Section 4 and the capabilities of RFID technology in Section 5, we can now turn our attention on how the latter can cater to the former. In doing so, we shall make a distinction between container and item-level tagging since extending the application of RFID to every product item has significant implications in terms of extra capabilities and applications that become possible but also far higher associated costs.

Handling efficiency. A clear benefit of RFID is that it allows for the fully automatic identification of products and containers without the need to preserve line-of-sight between reader and tag as is the case for bar codes. In Section 7, for example, we discussed in detail how RFID portals at the distribution centre docking bay doors are used to reduce manual data capture needs and expedite the delivery confirmation process. Similar scenarios can be developed for most shipping and receiving situations where container-level tagging would satisfy all logistics requirements. Further, item-level tagging would allow the development of additional consumer-facing applications, notably fast-scan point of sale portals that can considerably reduce queuing times for checkout.

Out-of-stock reduction. Despite the considerable progress of ECR and other such industry initiatives, stock-outs remain common at the retail store level. Case and pallet-level RFID tagging can increase product availability by reducing the number of delivery errors, by increasing inventory accuracy, and by improving the timely replenishment of products from the back store. Item-level tagging can further reduce stock-outs by providing precise information about inventory levels in the store front, rather than estimates based on sales data which can be erroneous in excess of ten per cent. Especially for clothing, item-level RFID can provide detailed information about the product mix that is actually available on the storefront shelves and thus significantly increase availability. This information can be captured either with portable readers and periodic inventory scans or by embedding readers in shelves as part of their construction. Although the latter approach is far more costly, it allows for the development of 'smart' shelves which also include additional small displays that expose additional inventory information, for example, products sizes that are available in the back room but not in the store front, an approach that has been proven especially successful for higher cost garments and shoes [26].

Inventory reduction. At this stage of RFID development, it is not possible to quantify or even confirm the possible potential of this technology to lower inventory levels for some or all trading partners, a fact which is especially critical in the case of FMCG. Although in some cases this would clearly be possible, for example, by helping avoid excess stock due to reorders of products already available but not immediately locatable in the storeroom, there is insufficient evidence that either container or item-level tagging will increase the accuracy or the timeliness of demand forecasts, which still appear to remain intractable. Similar reasoning applies to the case of unsaleables where traceability and common sense can play a role in reducing inventory levels. Further experience and research in this area is required although some estimates based on simulations employing simplified models of the FMCG supply-chain appear to be encouraging [24].

Order reconciliation. Container-level RFID can prevent delivery errors (as highlighted in the scenario of Section 7) and reduce the manual effort associated with delivery confirmation as well as the time required to complete this process from several minutes to almost immediate verification. This technology may also prove sufficient as a means of proof for shipment delivery and thus simplify dispute resolution. Nevertheless, in most cases disputes relate to pricing which is best addressed through the implementation of GDSN and RFID has little to offer in this respect. Item-level RFID does not appear to have any impact for order reconciliation.

Theft. While RFID at the container level does not prevent theft, it may assist with the detection of specific sections of the supply chain where this problem is of particular significance. Item-level RFID, on the other hand, can have considerable implications as it is already a proven technology for anti-theft systems for retail. Furthermore, item-level tagging makes far more challenging the re-introduction of stolen or indeed counterfeit products into the supply chain and for this reason it has attracted intense interest, especially in the case of medicines and medical supplies in general.

Nonetheless, the focused performance metrics discussed in the preceding paragraphs may not be the whole story. In addition to the specific new information sources afforded by RFID which can be related directly to quantifiable optimization effects, it is likely that the technology has a secondary role as the catalyst for change. Indeed, to capitalize on the data produced by RFID systems and gain a competitive advantage, it is not enough to simply implement the technology but also to be able to transform data into meaningful business information that can be acted upon. This requires advanced integrated information technology infrastructure across the enterprise including warehouse management and enterprise resource planning systems, but perhaps more importantly a reorganization of business processes and strategies. Such changes require a long-term commitment and considerable investment in human resources and can potentially completely transform the way business is conducted.

Hence, the decision or not to implement RFID may in many cases extend well beyond a simple automation decision into a business change [38]. In this case, the implication is that the decision to implement the technology is associated with significant business risk and requires very careful planning and execution. In view of that, the business that decides to be involved in such technology implementation must be convinced of its benefits and be able to implement such a programme of change. But with convincing evidence lacking in many cases, this risk cannot be justified only on the basis of a cost-benefit analysis. Nevertheless, it is characteristic that the pioneers of bar code, the previous generation of auto-identification technology, have emerged as the dominant corporations in their respective domains and it is likely that this will be repeated for those willing to take well-calculated risks.

The rationale for making adecisions is quite different between container-level and item-level implementations. The former would be completely related to benefits in the supply chain and would explore the issues that we have already discussed in this and previous sections. Item-level tagging, on the other hand, has quite different value proposition as its high cost cannot be justified today or indeed in the foreseeable future for all types of products although specific application of limited scope can be easily developed for higher cost items.

In fact, widespread item-level tagging for products irrespective of their price is unlikely to be justified on the grounds of supply chain needs alone. Instead, item-level RFID is valuable for a variety of consumer services and indeed this is the most promising area for investigations of this technology and offers the promise of the most likely return on investment. Applications of this type have already appeared and are gaining in popularity [26]. For a discussion of related service development using a variety of sensors in addition to RFID, refer to [15].

Nevertheless, extending supply chain technologies in this way has significant repercussions for consumers who become directly involved in the enterprise data processing pipeline. Services employing item-level RFID use personal data associated with individual consumers in intimate ways and that can be used to reconstruct their private activities at an unprecedented level of detail. Moreover, recent studies indicate that the implementation of this technology may transform the consumption experience in unpredictable ways.

9. Consumer Acceptance of Item-Level Applications

Recent research in item-level RFID retail applications has identified a generally positive stance by consumers, especially when considering situations within the store. Project MyGrocer was the first to explore opportunities to develop such applications by focusing primarily around the concept of the smart shopping cart [28]. The working assumption of this work was that each product sold in a supermarket is individually tagged. The MyGrocer cart was fitted with a RFID reader so that every time a product was placed in it, it would be scanned and its code retrieved. The cart also carried a wireless computer with a large touch screen display connected to the reader (cf. Figure 14).

The main application provided three distinct areas of functionality. The first would present a shopping list, that is, a list of item for purchase selected by the individual shopper. This list would be associated with the profile of a specific user and created using historical purchase data, which can be further edited manually via a web interface on the supermarket web site. Each time one of the products on this list

FIG. 14. MyGrocer shopping cart in action at the Atlantic supermarkets during system testing.

would be placed in the cart, the item would be crossed out to confirm that it has been picked. The second application displays a running list of items in the cart, their quantities, their cost and the total cost of all the products in the cart. Finally, a third application would display information related to the last item picked, for example, ingredients, directions for use, health warnings and so forth. The same area of the screen may also be used to display offers and promotions, or comparisons with similar products to the ones in the shopping list. Finally, use of the smart shopping cart allows rapid checkout as the products are already scanned and the total price directly calculated.

This in-store scenario developed around item-level RFID received a favourable response, with the main benefit perceived to be the improvement of the shopping experience, which was understood to be faster, easier and to offer better value for money. The features of the applications that proved most attractive to consumers during the trials were:

- constant awareness of the total cost of the shopping cart content, which offers the opportunity to accurately control spending during a shopping trip,
- access to complete and accurate descriptions of products including price, size, ingredients, suitability for particular uses and so forth,
- the ability to compare the value of similar products,
- the provision of personalized, targeted promotions that reflect the individual consumer profile in addition to the usual generic promotions as well as the fact that they could access all offers available in the specific supermarket at a single contact point,
- the proposed in-store navigation system, especially in the case of hypermarkets where orientation is particularly complex,
- the smart checkout and the ability to bypass queues and reduce waiting time.

However, not all comments were positive. Focus groups and survey findings highlighted the collection of detailed personalized purchase statistics by the retailer and collaborating service providers to be of concern, even though the participants were aware of the provisions (although not the practicalities) of the data protection act. Their negative reaction to data collection was triggered primarily after (eponymous) authentication during log in to the shopping cart when, after presenting their RFID-enabled loyalty card and entering their private credentials, they were presented with their personalized shopping list. Two issues were raised, both relating to the immediate recognition that for the construction of the list, their past purchase data has been recorded, preserved and processed.

This reaction was more pronounced when considered in the context of MyGrocer applications outside the physical space of the store. In following with the ideas of consumer VMI explored in the previous section and in an attempt to collect supply chain data as early on in the consumption process as possible, the project also developed two additional scenarios that provided shopping list and ordering facilities: 'on the go' employed a cell phone to place orders, and 'at home' enabled the automatic collection of items for replenishment using RFID readers embedded at several positions at the use residence. The latter scenario, in particular, was the main source of concern since private data, collected in the sheltered space of the home, would be delivered to commercial organizations without the explicit control of the consumer.

Indeed, even in the more acceptable case of the store scenario, the vast majority of participants did not trust the retail service provider or the provider of the infrastructure to protect their privacy, irrespective of whether it was a contractual obligation or not. Moreover, the collection of very detailed information about their purchases over an extended period of time raised concerns about the use of the data for purposes that they have not consented to. They were also concerned that such availability of data could reveal their habits or private behaviours, especially to third parties that would subsequently gain access to these data.

Another major concern related to the overall shopping experience was perceived to point towards a technology controlled, fully standardized life-style. Two issues interrelate on this point. On the one hand, participants rejected the claim that a software system could predict accurately their wishes just by collecting historical data and monitoring habitual purchases. Indeed, this aspect of the system appeared to be patronizing and overtly rationalized, but most importantly contrary to the experience of being human. In fact, the majority of participants discarded the possibility of a computer system that could successfully predict their wishes, while some of them went as far as becoming offended by this suggestion as they interpreted it as denying their free will. On the other hand, the participants of the study perceived that such a system promoted primarily the interests of the supplier, while the consumer only received marginal benefits.

This issue of directly verifiable consumer value, or rather the lack thereof, was one of the two fundamental reasons for rejecting the system as a whole. Yet, this was not an absolute rejection of the system as the majority of participants in the studies would consider its use if they would receive appropriate compensation for the loss of privacy that they experience. The main challenge they set for retailers was how to fairly and appropriately strike a balance between their and the consumer benefit. The second core challenge before the system could become acceptable was that of control, in the sense that users demanded control over its operation. The form that this feature would take depended on the circumstances of its use and could vary from the anonymous use of the smart shopping cart (at the loss of the personalized shopping list feature), or indeed an off button for the RFID recording at home.

More recently, Metro Supermarkets in Germany has developed its so-called Store of the Future, which investigates ideas very similar in spirit to those explored by the MyGrocer store scenario. Although this activity is much more extensive in scope and intends to provide a full technological validation of modern RFID supply chain technologies, user studies have revealed that the same issues of control are still critical for consumers [20]. Although in this case again, shoppers would be willing to negotiate a loss of privacy in exchange for extra value, they wish to have control over when and how this would occur.

10.   Privacy Implications of Item-Level Tagging

Item-level RFID can provide retailers with unique sources of information that can be employed for applications beyond supply chain management. Such applications may offer welcome new shopping facilities to consumers, but at the same time, they also make possible new ways to violate personal privacy. Moreover, attacks on privacy enabled by item-level RFID are not limited to the physical confines of the store, but to all purposes extend to any public space and even to the intimate space of the home. In these cases, the risk is not solely due to the use of RFID by the retailer but rather by third parties using the availability of the technology to mount independent attacks on consumers.

There are two main types of privacy attacks that can be developed capitalizing on the widespread availability of item-level RFID tagging. In *tracking* attacks, the actions of individuals are recorded through the observation of RFID tags associated with their person, and their future behaviours potentially inferred. For example, an RFID tag that remains embedded into an item of clothing long after its purchase can be used to identify its wearer wherever they go. *Information leaks* happen when personal or intimate information stored in RFID tags is revealed without the consent of its owner [14, Chapter 4]. For example, when personal details encoded in a tag are skimmed from an e-passport without the owner consent. Both types of attacks become particularly likely when item-level tags affixed or embedded in consumer goods are not removed at the point-of-sale, so that stored identifiers can be retrieved by unauthorized readers, recorded and processed without any visible indication to the user that this activity occurs.

A closer examination of tracking attacks identifies several distinct scenarios that become possible through item-level tagging [13]. For example, one of the earliest uses of RFID outside the supply chain that was explored during the development of the EPC system was in anti-theft applications. This is of particular relevance to items of small size but high value such as replacement razor blades, which are the most common target of shoplifting. In this scenario, smart shelves would monitor high value items placed on them, and in case where a relatively large number be suddenly removed, a camera would take a photograph as evidence against a potential thief. But in practice, it is hard to differentiate between lawful behavior and attempts to steal and as a result photographs were taken in many more cases than it was necessary. Although this may appear as a minor compromise of privacy it is nevertheless highly suggestive of the type of applications that are possible and how easy it is to develop applications using flawed heuristics.

Consumer privacy violations can be are examined in finer granularity in terms of specific threats, to pinpoint the many ways in which data analysis techniques, profile data, and the presence or absence of specific products can lead to violating ones' rights

[13]. As noted earlier, the widespread availability of RFID tagged products present opportunities for covert data collection in locations and situations without the consent of the consumer. Individuals associated with particular product item tags can in this way linked with visits to specific locations at specific times. Even more, if readers observe several locations, sequences of visits can be reconstructed and using simple inference techniques common behaviors, habits or routines can be discovered.

Simpler but equally effective uses of the technology are also possible: a consumer carrying a particular type of product can be identified and approached with a discriminatory intention for example, because they carry a particular book title. A related use of the technology but with different intent, would see the consumer being approached as a result of their possessing a particular item or brand which reveal their preferences. Identification of such preferences can be an effective marketing tool for competing retailers or simply used to identify the value of ones' property and identify them as a worthwhile target of criminal intentions.

Such techniques are more effective when considering constellations rather than single products. Depending on the fact that a particular person is singularly associated with a specific product item may be haphazard as products can be shared between several consumers, tracing collections of product identifiers moving together in a single constellation can provide much more accurate results. Even more so, when individual items are shifted from an established constellation into another, then it is possible to conclude that a transaction has taken place between the two persons involved.

Observing product items or product constellations over extended periods of time can provide adequate information to predict or infer preference or behaviors. Although this is to some extend possible today through the use of loyalty schemes and cell phone records, tracking RFID tags does not require a contractual relationship with the consumer due to the technical characteristics of RFID. Moreover, RFID readers can be installed in such a way that there is no perceptible indication of their existence. Even when data collection in this way is carried out within the provisions of a mutually agreed upon contract the wealth of information collected makes the indirectly enforced use of the technology through preferential pricing particularly attractive can significantly reduce the capability of consumers to make free choices.

Last but not least, RFID tags can be used as a physical equivalent of cookies with the vast majority of preferential pricing techniques developed for the web directly applicable [1]. Indeed, historical information about acceptance or rejection of offers or other transaction opportunities can be stored on one or more tags carried by an individual and used to tailor future approaches to fit their profile. This is certainly feasible for the retailer that supplier the particular item used as carrier but due to the generally inadequate security provisions of RFID, this technique could well be accessible to third parties.

RFID technology has been the cause of the majority of privacy concerns, early commercial applications have not helped to develop public confidence as many events show. For example, Metro Supermarkets in Germany violate their own stated privacy policy by embedding covert RFID tags in their loyalty cards, and an early briefing of Auto-ID Center sponsors urged them to capitalize on consumer apathy and push for item-level tagging thus creating a de-facto situation before consumer organizations could react [9].

The lack of adequate security and privacy protection provisions has sparked intense interest in this area of RFID technology. It is characteristic that during the year 2006 almost a century of research papers have been published in this area and the trend is accelerating. Despite this fact, the main RFID standards have already been ratified without adequate provisions and during the time of publication of these research over 1.5 billion tags entered circulation.

11. RFID and EU Law

Although the consensus appears to be that RFID is a critical technology for future economic growth across several industrial sectors, it is also clear that its application must also be socially and politically acceptable, ethically admissible and legally allowable. This aim becomes even more complex to achieve due to the universal scope of RFID technology, which must respect the policies, ethics and law of every region and country where it is employed. To be sure, this is a challenging task and in an attempt to make the main issues tractable from a computing perspective, in this section we will discuss the main considerations as they relate to the legal framework of the European Union.

11.1 Data Protection and Privacy

The EU founding treaty declares the fundamental freedoms that its citizens may expect including liberty, democracy and respect for human rights. Article 30 of the treaty in particular requires the enforcement of appropriate provisions for the protection of personal data including the collection, storage, processing, analysis and exchange of information. Moreover, Article 8 of its Charter of Fundamental Rights proclaims the protection of personal data as one of the freedoms that each citizen has a right to enjoy.

These principles are interpreted and implemented in practice through the legislative framework for data protection and privacy. The Data Protection directive in particular has been developed aiming to provide the general rules and the long-term vision and to be robust despite technological innovations. Privacy protection is specifically addressed within the directive and is expressed in a way that is independent of the specific techniques and mechanisms employed in information processing and thus also applies in the case of RFID.

This directive is complemented by the more recent Privacy and Electronic Communications directive (also known as ePrivacy directive). This extension applies the general principles to the processing of personal data for the provision of public electronic communications services over public communications networks as well as to the recording and use of location data. It also specifies that direct marketing communications are only allowed when the recipient has agreed to be contacted in advance or in the context of an existing customer relationship, in which case companies can continue to market their own similar products on an opt-out basis. However, since RFID in most cases operates over private or corporate networks, it has been argued that the provisions of the ePrivacy directive do not apply although this is only one interpretation which does not take into account the case of the use of RFID readers in public spaces.

11.2   Commercial Transactions

The Electronic Commerce directive regulates the process of contract offer and acceptance and applies to the fast checkout process supported by RFID points of sale. The eCommerce directive has several provisions regarding appropriate ways of notifications of contractual terms and conditions and dictates that explicit consumer consent be given at all stages. Although exceptions apply to cases in which the interaction medium does not allow for information-rich interactions, RFID's predominantly silent operation stresses this requirement to its limit.

11.3   Governance

A central issue that affects the implementation of RFID is governance in the sense of access to RFID-related standards and infrastructures. The EU has been conceived as a vehicle for economic collaboration and has a tradition of creating a common open and non-discriminatory set of rules which strive to promote fairness and interoperable infrastructures. As such, its regulating bodies take a particularly negative view of any attempt to fragment public or shared infrastructures or the deployment of proprietary systems with the specific objective to prevent competitors from entering a market.

Arguably, the EPC system is tightly controlled by a group of companies and developed with a view to serve their interests and specific ends which relate to commercial, security and political aspects of governance. Furthermore, the spirit of the community is one where protection is not limited to individuals but extends to companies, whose sensitive commercial information is also protected as is the case of data within RFID-enabled business processes. As such, it is natural to expect the two opposite sides of the EPC proposition, namely, rapid development of a new market sector and proprietary technology and infrastructures, will cause considerable friction and can potentially lead to closer regulation.

11.4   Spectrum Regulation

Recently, the EU has opted to liberate more spectrum for the growing demand for RFID usage, implemented through Decision 12 for RFID frequencies in the UHF band adopted by the Commission. This establishes a harmonized base for RFID applications across European states but nevertheless does not completely address the problem. In some cases, for example in distribution centres and shopping malls, it is necessary to operate hundreds or even thousands of readers in close proximity to each other in event driven mode. However, ETSI, the European Telecommunications Standards Institute, in standard EN 302 208 requires the use of Listen Before Talk to prevent a base station from transmitting if the channel is already occupied by another transmission. This limits the number of readers able to operate simultaneously in a particular radio neighborhood to about twenty if all available channels are used and has some incompatibilities with Gen2 tag operation.

11.5   Environmental Issues

There are two directives that have already had very significant repercussions for electronics in general and RFID in particular, namely, on waste electrical and electronic equipment (WEEE) and on the restriction of the use of certain hazardous substances in electrical and electronic equipment (RoHS). RoHS in particular bans the use of certain hazardous substances which are rather common in electronics.

Relating to public health, the EU has some of the most strict regulation of the level of electromagnetic regulations that workers or the general public may be exposed to. Moreover, the Commission has in place a regular program of monitoring the possible effects of electromagnetic fields on human health through its Scientific Committees. Moreover, restrictions on EMF emissions from products available in any European state have been established to ensure the safety of both users and non-users. Although electromagnetic fields created by RFID equipment are generally low and thus exposure of the general public and workers is expected to be well below current limits, RFID nevertheless contributes to the total radiation in working and home environments and its widespread use may well have significant results, especially when taking into account wireless networking technologies used in tandem.

12. Discussion and Conclusions

Many believe that technology and business dominate culture today, yet it is a society's privacy culture that defines its values, sensibilities and commitments. To be sure, attitudes toward privacy change as technologies that blur the distinctions between what is public and private emerge. Deploying any new technology involves risk, and society relies on experts to accurately assess that risk; failure to do so compromises their role as gatekeepers. It is thus the responsibility of the computing profession to confront the challenges of RFID in retail. How we deal with these issues will determine the chances of widespread adoption of not only RFID but potentially the whole range of emerging ubiquitous computing technologies.

Advising that deployment of RFID, or any technology for that matter, should exploit 'consumer apathy' does little to inspire public trust, as does making a tag impossible to remove. Two aspects of the technology accentuate the trust problem and dictate collaboration across disciplines:

- RFID-based systems' silent and transparent operation; and
- the fact that trust is not a purely cognitive process and thus is not amenable to a strictly quantitative treatment, for example, as a personal utility optimization problem, a popular view within computer science today.

In fact, many of the core challenges involve managing the enormous amounts of data that RFID generates and monitoring the massive increase in points of contact between user and system rather than developing cryptographic algorithms and security mechanisms that control access to tag data. While individuals' initial entitlement to control their data is well recognized, economic coercion mechanisms based on price discrimination are less so. Such mechanisms result from negotiations between private organizations and public institutions, and this is where our professional social responsibility must play a critical role. Dealing effectively with misuse will become more urgent in the near future.

This survey has attempted to provide an in-depth description of issues and technologies and to supply computing professionals with the information needed. Yet, several issues related to large-scale deployments of RFID are still poorly understood and we could not conclude this discussion without exploring the additional implications from a waste management perspective caused by the extensive use of RFID.

Indeed, RFID tags routinely embedded in a variety of products affect a wide gamut of recycling processes, both of materials used in containers for the supply and in product item packaging. For example, as relates to paper recycling adhesives, chips, pieces of metal from antennae and conductive inks affect the process of reclaiming containers and paperboard and prevent the manufacture of new board from recycled feedstock.

Similar effects would be caused contamination on steel, glass and plastic recycling processes by RFID tag debris. Furthermore, at the end of their useful life, pallets are ground up for use as landscape mulch, animal bedding, compost, soil amendment, or core material for particle board. However, metallic pieces from antennae will be shredded, but cannot break down and would pollute the composting process and render the material unusable. It is ironic that RFID is often seen as the solution for reclaiming materials from consumer products due to its capability to record an accurate and complete history of the product. At least in the short term, its effect will almost certainly be negative.

REFERENCES

[1] Acquisti A., 2006. Ubiquitous computing, customer tracking, and price discrimination, in Roussos G., ed. Ubiquitous and Pervasive Commerce (Springer, London), pp. 115–132.

[2] Bornhövd C., Lin T., Haller S., and Schaper J., 2004. Integrating automatic data acquisition with business processes – experiences with SAP's auto-ID infrastructure, in *Proc. VLDB04.*

[3] Brown S. A., 1997. *Revolution at the Checkout Counter: The Explosion of the Bar Code* (Harvard University Press).

[4] Caneel R., and Chen P., 2006. Enterprise Architecture for RFID and Sensor Based Services (Oracle Corporation, Redwood Shores).

[5] Gemini C., 2005. Global Data Synchronisation At Work in the Real World: Illustrating the Business Benefits (Global Commerce Initiative).

[6] Chamberlain J., Blanchard C., Burlingame S., Chandramohan S., Forestier E., Griffith G., Mazzara M. L., Musti S., Son S-I., Stump G., and Weiss C., 2006. IBM WebSphere RFID Handbook: A Solution Guide (IBM Redbooks, Raleigh).

[7] Chen H., Chou P. B., Duri S., Elliott J. G., Reason J. M., and Wong D. C., 2005. A model-driven approach to RFID application programming and infrastructure management, in *Proc. ICEBE05 (IEEE Press)*, pp. 256—259.

[8] Curty J-P., Declercq M., Dehollain C., and Joehl N., 2006. Design and Optimization of Passive UHF RFID Systems (Springer, Berlin).

[9] Dunne H., June 2002. Message Development, Auto-ID Sponsor briefing.

[10] Ferry G., 2003. A Computer Called LEO: Lyons Teashops and the World's First Office Computer, Fourth Estate.

[11] Finkenzeller K., 2003. RFID Handbook: Fundamentals and Applications in Contactless Smart Cards and Identification (John Wiley & Sons, London).

[12] Floerkemeier C., and Lampe M., 2005. RFID middleware design – addressing application requirements and RFID constraints, in *Proc. SOC-EUSAI, ACM International Conference Proceeding Series*, Vol. **121**, pp. 219–224.

[13] Garfinkel S. L., Juels A., and Pappu R., 2005. RFID Privacy: an overview of problems and proposed solutions. *IEEE Security and Privacy*, 3(3):34–43.

[14] Garfinkel S., and Rosenberg B., 2005. RFID: Applications, Security, and Privacy (Addison-Wesley).

[15] Gershman A., and Fano A. E., 2003. Customer service with Eyes, in *Proc. Work. Ubiq. Comm. (electronic proceedings)*.

[16] Global Commerce Initiative, An Integrated View of the Global Data Synchronisation Network and the Electronic Product Code Network (IBM Consulting Services, 2004).

[17] Grocery Manufacturers of America, Full-Shelf Satisfaction: Reducing Out-of-Stocks in the Grocery Channel (2002).

[18] Grocery Manufacturers of America, A Balanced Perspective: EPC/RFID Implementation in the CPG Industry (2004).

[19] Gruen T. W., Corsten D. S., and Bharadwaj S., 2002. Retail Out-of-Stocks: A Worldwide Examination of Extent, Causes and Consumer Responses (Grocery Manufacturers of America, The Food Marketing Institute).

[20] Günther O., and Spiekermann S., 2005. RFID and the perception of control: the consumer's view, *Comm. ACM*, **48**(9):73–76.

[21] Hoag J. E., and Thompson C. W., 2006. Architecting RFID Middleware, *IEEE Int. Comp.* **10**(5):88–92.

[22] Hollinger R. C., and Davis J. L., 2002. National Retail Security Survey 2001 (Center for Studies in Criminology and Law, University of Florida).

[23] Jeffery S. R., Garofalakis M., and Franklin M. J., 2005. Adaptive cleaning for RFID data streams, in *Proc. VLDB05*, pp. 163–174.

[24] Kelepouris T., Pramatari K., and Doukidis G., 2007. *RFID-enabled traceability in the food supply chain, Ind. Man. & Data Systems*, **107**(2):183–200.

[25] Kim Y., Moon M., and Yeom K., 2006. A Framework for Rapid Development of RFID Applications, in *Proc. ICCSA 2006, Lecture Notes in Computer Science*, Vol. **3983**, pp. 226–235.

[26] Konomi S., and Roussos G., 2007. Ubiquitous computing in the real world: lessons learnt from large scale RFID deployments, *Pers. Ubiq. Comp.*, forthcoming.

[27] Kotok A., and Webber D. R. R., 2001. ebXML: The New Global Standard for Doing Business on the Internet (New Riders Publishing).

[28] Kourouthanassis P., and Roussos G., 2003. Developing consumer-friendly pervasive retail systems, *IEEE Perv. Comp.*, **2**(2):32–39.

[29] Landt J., 2005. The history of RFID, *IEEE Potentials*, **24**(4):8–11.

[30] Lightburn A., 2002. Unsaleables Benchmark Report, Joint Industry Unsaleables Steering Committee (Food Marketing Institute and Grocery Manufacturers of America).

[31] Martin A. J., 1995. Infopartnering: The Ultimate Strategy for Achieving Efficient Consumer Response (John Wiley & Sons).

[32] McGuffog T., and Wadsley N., 1999. The general principles of value chain management, Supp. Ch. Man, **4**(5):218–225.

[33] Mealling M., 2002. Dynamic Delegation Discovery System (DDDS) Part Three: The Domain Name System (DNS) Database (IETF).

[34] Prabhu B. S., Su X., Ramamurthy H., Chu C.-C., and Gadh R., 2006. WinRFID A Middleware for the enablement of radio frequency identification (RFID) based applications, in Shorey R., Choon C. M., Tsang O. W., and Ananda A., eds., *Mobile, Wireless and Sensor Networks: Technology, Applications and Future Directions* (Wiley-IEEE Press).

[35] Raman A., 2000. Retail-Data Quality: Evidence, causes, costs, and fixes, *Tech. Soc.*, **22**(1):97–109.

[36] Römer K., Schoch T., Mattern F., and Dübendorfer T., 2004. Smart identification frameworks for ubiquitous computing applications, Wireless Networks **10**(6):689–700.

[37] Roussos G., and Moussouri T., 2004. Consumer Perceptions of Privacy, Security and Trust in Ubiquitous Commerce, *Pers. Ubiq. Comp.*, **8**(6):416–429.

[38] Roussos G., 2006. Enabling RFID in retail, *IEEE Computer.*, **39**(3):25–30.

[39] Roussos G., 2005. Ubiquitous and Pervasive Commerce: New Frontiers for Electronic Business (Springer, London).

[40] Smaros J., and Holmstrom J., 2000. Reaching the consumer through e-grocery VMI, Int. J. Retail Distr. Man., **28**(2):55–61.

[41] Tellkamp, 2006. The impact of auto-ID technology on process performance RFID in the FMCG supply chain, Technical Report (Auto-ID Lab St. Gallen).

[42] Vogt H., 2002. Efficient object identification with passive RFID tags, in *Proc. Pervasive 2002, Lecture Notes Computer Science*, Vol. **2414**, pp. 98–113.

[43] Wang F., and Liu P., 2005. Temporal management of RFID data, in *Proc. VLDB05*, pp. 1128–1139.

Medical Robotics and Computer-Integrated Interventional Medicine*

RUSSELL H. TAYLOR AND PETER KAZANZIDES

Department of Computer Science
The Johns Hopkins University
Baltimore, Maryland

Abstract

This chapter is concerned with computer-integrated interventional medicine (CIIM). The fundamental premise of CIIM, which is sometimes also referred to as Computer-Integrated Surgery (CIS), is that the use of information and information-driven systems will fundamentally change clinical care by improving physicians' ability to plan, execute and follow up surgical and other interventional procedures. We first introduce the concepts of Surgical (or Interventional) CAD/CAM and Surgical Assistance. Next, we discuss the basic technology and techniques underlying these systems and provide examples of typical Surgical CAD/CAM and Surgical Assistant Systems. We conclude with a few thoughts about the growing significance and future prospects of the field.

  * Reprinted from *Biomedical Information Technology*, D. Feng, Ed.; Russell Taylor and Peter Kazanzides, "Medical Robotics and Computer-Integrated Interventional Medicine", pp. 393–416, 2007, Reprinted with permission from Elsevier.

219

1. Introduction

This chapter is concerned with computer-integrated interventional medicine (CIIM). Over the past 50 years, the technology used in interventional medicine increasingly has been computer-based. Medical imaging devices have progressed from simple x-ray units to sophisticated systems, combining advanced sensors and computation to provide unprecedented information about a patient's anatomy and physiology. Medical workstations are able to combine information from many sources to help surgeons and other physicians to plan interventions and to provide real-time information supports in carrying out these plans. Robotic devices and endoscopic cameras enable physicians to perform minimally invasive procedures that would otherwise be impossible. Computer-controlled systems use directed energy to destroy tumours and other malformations inside a patient's body without surgery. Computer-based physiological monitoring devices are ubiquitous in operating rooms and intensive care units.

This evolution is a natural consequence of the computer's ability to integrate information with action to fundamentally improve treatment processes, in much the same way that computer-integrated systems and processes have affected other sectors of our society, such as manufacturing, transportation, retailing and agriculture. The basic 'information loop' of interventional medicine is illustrated in Fig. 1. The process starts with information about the patient, such as images, lab results, genetic information and symptoms. This information is combined with general information about human

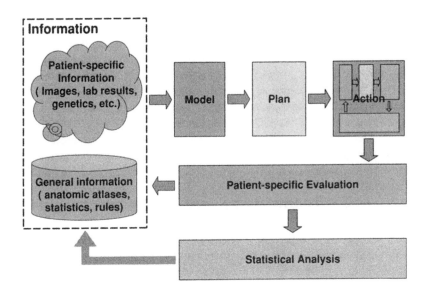

FIG. 1. Computer-integrated interventional medicine (CIIM) as a closed-loop process.

anatomy and physiology to create a patient-specific 'model' or representation that is used to diagnose the patient's condition and formulate an interventional plan. During the intervention, the 'virtual reality' of the model and plan are registered to the 'actual reality' of the patient and may be coupled with appropriate technology to assist the clinician in carrying out the plan. Further information is typically generated both during and after the intervention to update the model and to assess the effect of the intervention. This information may be used subsequently in further treatment of the patient. It may also be analysed statistically to assess and improve the overall effectiveness of treatment plans and protocols, in a manner somewhat analogous to the use of statistical quality control and process learning in manufacturing.

We often refer to this closed-loop process of i) constructing a patient-specific model and interventional plan; ii) registering the model and plan to the patient; iii) using technology to assist in carrying out the plan; and iv) assessing the result as *Surgical (or Interventional[1]) CAD/CAM*, again emphasizing the analogy between computer-integrated interventional medicine and computer-integrated manufacturing. Of course, it is important to recognize that there are also profound differences

---

[1] The terms *Computer-Integrated Surgery* and *Computer-Assisted Surgery* have often been applied to the concepts discussed in this chapter. However, the use of the word 'surgery' is somewhat limiting and causes a certain amount of discomfort among interventional radiologists and others who perform computer-assisted or image-guided interventions but who do not consider themselves to be 'surgeons'. Consequently, we will frequently use the more general terms 'interventional medicine' and 'interventionist'.

between medicine and manufacturing. In particular, our goal is *not* automation of medical interventions. Rather, our goal is to exploit computer-based technology and systems to assist human clinicians in treating patients. Thus, we often refer to these systems as *Surgical (or Interventional) Assistants*, especially when the interventional decisions are highly interactive, as is frequently the case with surgery. However, it is important to remember that these concepts are not incompatible. Although it is often more convenient to think of a CIIM system as being primarily a CAD/CAM or an Assistant system, the same underlying concepts and technology are present in both cases. As these systems become more and more sophisticated, the distinction will be harder and harder to make.

2. Technology and Techniques

In this section, we will provide a brief overview of key technology components found in CIIM systems, with special attention to surgical navigation and medical robotics. Further discussion may be found in [1–4].

2.1 System Architecture

The overall architecture of CIIM systems is shown in Fig. 2. Broadly, these systems consist of the following components: 1) Computational components performing a wide variety of image processing, surgical planning, monitoring and similar tasks; 2) databases of patient-specific information, as well as more generic knowledge bases about human anatomy and physiology, common treatment plans, outcome data, etc., and 3) devices such as images, robots, and human–machine interfaces relating the 'virtual reality' of computer representations to the 'actual reality' of the patient, interventional room and clinician.

2.2 Registration and Transformations Between Coordinate Systems

Geometric relationships between portions of the patient's anatomy, images, robots, sensors and equipment in interventional suites are fundamental in all areas of computer-integrated interventional medicine, and there is an extensive literature on techniques for determining the transformations between the associated coordinate systems (e.g., [5, 6]). The brief discussion below follows the basic framework developed in [6]. Given two coordinates $\vec{v}_A = [x_A, y_A, z_A]$ and $\vec{v}_B = [x_B, y_B, z_B]$ corresponding to comparable features in two coordinate systems Ref$_A$ and Ref$_B$, the process of registration is simply that of finding a function $\mathbf{T}_{AB}(\cdots)$ such that

$$\vec{v}_B = \mathbf{T}_{AB}(\vec{v}_A)$$

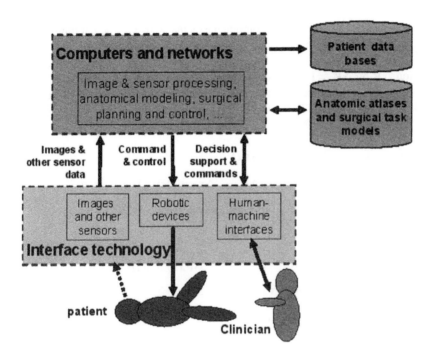

FIG. 2. The architecture of computer-integrated interventional medicine systems.

Although non-rigid registrations are becoming more common, $T_{AB}(\cdots)$ is still usually a rigid transformation of the form

$$\vec{v}_B = \mathbf{T}_{AB}(\vec{v}_A) = \mathbf{R}_{AB} \cdot \vec{v}_A + \vec{p}_{AB}$$

where $\mathbf{R}_{AB}$ represents a rotation and $\vec{p}_{AB}$ represents a translation. $\mathbf{R}_{AB}$ is often represented by an axis $\vec{n}$ and angle θ so that

$$\mathbf{R}_{AB}(\vec{n}, \theta) = e^{\theta \hat{n}} \quad \text{where } \hat{n} = \begin{bmatrix} 0 & -n_z & n_y \\ n_z & 0 & -n_x \\ -n_y & n_x & 0 \end{bmatrix}$$

Thus, if we have two transformations $\mathbf{T}_{AB}$ and $\mathbf{T}_{BC}$, the rotation and displacement components associated with the composite transformation will be given by

$$\mathbf{R}_{AC} = \mathbf{R}_{AB} \cdot \mathbf{R}_{BC}$$
$$\vec{p}_{AC} = \mathbf{R}_{AB} \cdot \vec{p}_{BC} + \vec{p}_{AB}$$

In many cases, $\mathbf{T}_{AB}$ cannot be computed exactly, so that the actual transformation $\mathbf{T}^*_{AB}$ is related to the nominal value $\mathbf{T}_{AB}$ by a small perturbation, i.e., $\mathbf{T}^*_{AB} = \mathbf{T}_{AB} \cdot \Delta \mathbf{T}_{AB}$. In this case, we frequently approximate the rotational component of a small rotation $\Delta \mathbf{R}$ by $\Delta \mathbf{R} \approx \mathbf{I} + \theta \hat{\mathbf{n}}$, so that $\Delta \mathbf{R} \cdot \vec{\mathbf{v}} \approx \vec{\mathbf{v}} + \theta \hat{\mathbf{n}} \times \vec{\mathbf{v}}$. Furthermore, we often ignore the effects of a small rotation $\Delta \mathbf{R}$ on a sufficiently small translation vector $\Delta \vec{\mathbf{p}}$, so that $\Delta \mathbf{R} \cdot \Delta \vec{\mathbf{p}} \approx \Delta \vec{\mathbf{p}}$. Thus, if the actual value of a coordinate $\vec{\mathbf{v}}^*_A \approx \vec{\mathbf{v}}_A + \Delta \vec{\mathbf{v}}_A$, then the actual value of $\vec{\mathbf{v}}^*_B = \mathbf{T}^*_{AB} \cdot \vec{\mathbf{v}}^*_A$ will be given by

$$
\begin{aligned}
\vec{\mathbf{v}}^*_B &= \mathbf{T}_{AB} \cdot \Delta \mathbf{T}_{AB} \cdot \left(\vec{\mathbf{v}}_A + \Delta \vec{\mathbf{v}}_A\right) \\
&= \mathbf{T}_{AB} \cdot \left(\Delta \mathbf{R}_{AB} \cdot \vec{\mathbf{v}}_A + \Delta \mathbf{R}_{AB} \cdot \Delta \vec{\mathbf{v}}_A + \Delta \vec{\mathbf{p}}_{AB}\right) \\
&\approx \mathbf{T}_{AB} \cdot \left(\vec{\mathbf{v}}_A + \theta \hat{\mathbf{n}} \times \vec{\mathbf{v}}_A + \Delta \vec{\mathbf{v}}_A + \theta \hat{\mathbf{n}} \times \Delta \vec{\mathbf{v}}_A + \Delta \vec{\mathbf{p}}_{AB}\right) \\
&\approx \mathbf{T}_{AB} \cdot \left(\vec{\mathbf{v}}_A + \theta \hat{\mathbf{n}} \times \vec{\mathbf{v}}_A + \Delta \vec{\mathbf{v}}_A + \Delta \vec{\mathbf{p}}_{AB}\right) \\
&= \mathbf{R}_{AB} \cdot \left(\vec{\mathbf{v}}_A + \theta \hat{\mathbf{n}} \times \vec{\mathbf{v}}_A + \Delta \vec{\mathbf{v}}_A + \Delta \vec{\mathbf{p}}_{AB}\right) + \vec{\mathbf{p}}_{AB} \\
&= \vec{\mathbf{v}}_B + \mathbf{R}_{AB} \cdot \left(\theta \hat{\mathbf{n}} \times \vec{\mathbf{v}}_A + \Delta \vec{\mathbf{v}}_A + \Delta \vec{\mathbf{p}}_{AB}\right)
\end{aligned}
$$

Thus, the uncertainty in $\vec{\mathbf{v}}_B$ will be given by

$$
\Delta \vec{\mathbf{v}}_B = \mathbf{R}_{AB} \cdot \left(\theta \hat{\mathbf{n}} \times \vec{\mathbf{v}}_A + \Delta \vec{\mathbf{v}}_A + \Delta \vec{\mathbf{p}}_{AB}\right).
$$

There is an extensive literature concerning registration methods. Typically, the process involves finding corresponding sets of features F_A and F_B and then finding a transformation $\mathbf{T}_{AB}(\cdots)$ that minimizes some distance function $d_{AB} = distance(F_B, \mathbf{T}_{AB}(F_A))$. Typical features can include artificial fiducial objects (pins, implanted spheres, rods, etc.) or anatomical features such as point landmarks, ridge curves or surfaces. One very common case involves registration of a set of sample points from an anatomical surface with a computer representation of that surface. In this case, variations of the Iterated Closest Point algorithm of Besl and McKay [7] are commonly used. For example, 3-D robot coordinates $\vec{\mathbf{a}}_j$ may be found for a collection of points known to be on the surface of an anatomical structure which can also be found in a segmented 3D image. Given an estimate $\mathbf{T}_k$ of the transformation between image and robot coordinates, the method iteratively finds corresponding points $\vec{\mathbf{b}}^{(k)}_j$ on the surface that are closest to $\mathbf{T}_k \cdot \vec{\mathbf{a}}_j$ and then finds a new transformation

$$
\mathbf{T}_{k+1} = \arg \min_{\mathbf{T}} \sum_j \left\| \vec{\mathbf{b}}^{(k)}_j - \mathbf{T}_k \cdot \vec{\mathbf{a}}_j \right\|^2
$$

The process is repeated until some suitable termination condition is reached.

2.3 Navigational Trackers

Real-time measurement of intraoperative positions and orientations is ubiquitous in CIIM, and a number of different technologies are available for this purpose. These include encoded mechanical linkages, ultrasound localizers, electromagnetic localizers, 'active' optical triangulation systems that locate light emitting diodes, 'passive' optical triangulation systems that locate reflective markers and more general computer vision systems. Excellent technology surveys may be found in books such as [8–10] and in papers comparing different systems (e.g., [11–13]), although one should be aware that the relative technical capabilities of different technology approaches can change as technology develops.

In recent years, optical systems such as the Optotrak® and Polaris® systems (Northern Digital, Inc., Waterloo, Canada) have been the most widely used option for surgical navigation systems (see Section 4.2 and Fig. 10) because of their relatively high accuracy, predictable performance, and insensitivity to environmental variations. However, they do have several limitations. The most serious of these is the requirement that a clear line of sight be maintained between the tracking cameras and the markers being tracked, which can complicate the arrangement of equipment and workflow around the patient. A related drawback is that the markers being tracked must generally be on portions of surgical instruments outside the patient. This approach can lead to inaccuracies in instrument tip position determination and cannot be used with flexible instruments such as catheters.

Electromagnetic trackers were considered for many early surgical navigation applications, but the measurement distortions associated with metal in operating rooms caused them to fall out of favour. More recently, improvements in electromagnetic tracking technology (including reduced distortion and the development of very small sensors) and increased interest in tracking devices inside the patient has led to increased interest in this technology. Current examples include the Aurora® (Northern Digital, Waterloo, Canada), Flock-of-Birds® (Ascension Technology, Burlington, Vermont), Polhemus Patriot (Polhemus, Inc., Burlington, Vermont) and proprietary systems used in the Medtronic Axiem® (Medtronic Navigation, Inc. Louisville, Colorado) and the GE InstaTrak® (General Electric OEC Medical Systems, Salt Lake City, Utah).

2.4 Robotic Devices

Historically, the term *robot* has been used for multi-axis machines that are capable of autonomous motion. With this strict definition, the well-known daVinci system would not be classified as a robot, but rather as a *teleoperator*, because it does not operate autonomously. In fact, this would be true of many of the medical robot systems that have been developed in recent years. Therefore, at least in the medical field, the definition of a robot has been expanded to include virtually any mechanism that provides assistance to the surgeon, whether or not it can operate autonomously. In fact, safety is such a critical concern in medical robotics that it has prompted several researchers to develop robots that are incapable of autonomous motion (e.g., [14–17]). These systems rely on the surgeon, rather than motors, to provide sufficient force to create motion. The systems may still contain powered elements (e.g., motors, brakes), but they are only used to constrain motion. Although such systems do not fit the classical definition of a robot, they are considered *passive robots* in the medical field.

In an industrial setting, the benefit of robotics over fixed automation is that a robot can be programmed to serve in many different capacities. An industrial robot can assemble typewriters, weld car bodies or debur molded parts. There is, of course, some degree of specialization. A robot that places surface mount components on a printed circuit board is likely to be small and extremely accurate, whereas a robot that installs automobile windshields must be large and powerful. This specialization also applies to medical robots. For example, a robot developed for microsurgery will differ from a robot developed for orthopaedic joint reconstruction. Although the field of medical robotics is not yet mature, current experience suggests that medical robots may be more specialized than their industrial counterparts; a robot developed for one medical procedure may not be as easily adapted for other procedures, for reasons outlined below. Some examples of multi-functional medical robots do exist, such as the orthopaedic robot systems that assist with hip and knee replacement surgery as well as ligament repair.

Clearly, there are many similarities between industrial and medical robots: both (typically) consist of motors, sensors and articulated links that can be programmed to perform a variety of functions. There are, however, many differences, including the integration in the working environment, the relationship between the robot and the workpiece, the mechanical design and the strategies for assuring safe operation. These issues are discussed in the following paragraphs.

Robots are now commonplace on factory floors, and much experience has been gained in workcell configuration. Workcell design is simplified by the fact that other pieces of equipment, such as conveyer belts and parts feeders, are designed to integrate with robots and other industrial machines. Once a workcell design is completed, the robot and associated equipment are installed and, in most cases, left in place for a long time. In contrast, robots are not (yet) standard equipment in the interventional suite or operating room, where space is limited. Thus, a medical robot must be easily transported in and out of the room or, if permanently installed, should be able to be moved out of the way. In effect, a medical robot must be 'installed' in the *medical workcell* for each use. This installation includes transporting the robot to the site (e.g., operating room), connecting it to appropriate power sources, and sterilizing it.

Since other medical equipment is not designed for compatibility with robotics, the robot must 'fit in' as unobtrusively as possible. It is important to minimize the space requirement around the operating table, since much of this space is needed for the medical team and equipment.

Similarly, the manufacturing industry has widely adopted the principle of 'design for manufacturability', which means that parts are designed for ease of manufacturing by automated machines, including robots. Furthermore, in an industrial setting, the number of distinct parts is limited and like parts typically differ only by small manufacturing tolerances. The environment can be further structured using specialized parts feeders to orient or align the parts. In contrast, medical robots must operate on human anatomy, which cannot be redesigned to facilitate robotic procedures and is often not easily accessible from outside the body. Also, although humans have the same types of parts, there are large variations between individuals. A medical robot must be able to sense and adapt to these variations. If sensors alone cannot perform this task (and they often cannot), the clinician should be included 'in the loop' to augment the system's sensing capabilities. This requires a human–machine interface that is easy to use by individuals (clinicians) who do not have robotics backgrounds. In addition, novel kinematic designs are often necessary to be able to operate on the target anatomy without unduly restricting the clinician.

Although the mechanical design of medical robots has many similarities to that of industrial robots, the special requirements associated with interventional procedures (access, workspace, biocompatibility, imaging-device compatibility, etc.) has tended to produce distinct designs. For example, many medical robots are designed to manipulate surgical instruments or needles passed through constrained entry points into the patient's body. This consideration has led many groups (e.g., [17–20]) to develop kinematic structures that decouple tool orientation motions about a 'remote center of motion (RCM)' distal to the robot's structure. In clinical use, the robot is typically positioned so that the RCM point is positioned at the point where the instrument or needle passes into the patient's body (see Fig. 5 for an example). Similarly, a number of groups (e.g., [21–27]) have developed robots specifically for use in an MRI imaging environment.

Safety is an important consideration for both industrial and medical robots (e.g., [28]). In both cases, the goals, in order of priority, are: 1) to prevent injury to human beings working near the robot, and 2) to prevent the robot from damaging itself, other equipment, or the workpiece. In an industrial setting, safety systems typically involve gates, pressure-sensitive mats, and flashing lights – devices designed to keep people out of the robot's workspace or to shut down the system if a person comes too close. This is especially important when the robot is capable of high speeds or torques. In an industrial robot, high speeds and torques are desirable because they reduce the cycle time, thereby increasing the robot's productivity. In addition, many industrial robots require super-human strength to perform their tasks (e.g., lifting heavy parts). Unfortunately, these desirable attributes increase the potential danger to human beings. In the medical domain, there is little distinction between the two safety goals listed above since the 'workpiece' is a human patient, and 'other equipment' includes life-sustaining medical equipment. Because the medical staff and the patient must be inside the workspace, medical robot safety systems must ensure that they are not harmed even in the event of a malfunction. The situation is even more challenging for cases where the robot is holding a potentially dangerous device, such as a cutting instrument, and is supposed to actually contact the patient with this device (in the correct place, of course). As a result, compared with industrial robots, medical robots usually contain more redundancy in hardware and software and often have lower maximum speeds and torques.

Many classifications systems for medical robotics have been proposed [29]; some of them define systems as being *active*, *semi-active*, or *passive*. There is no universally accepted definition of these terms – some would argue that any robot that is capable of motion (i.e., contains powered actuators) can never be considered passive, whereas others focus on the manner in which the robot is used. This chapter adopts the latter convention, which is an operational definition rather than a mechanical definition:

- *Active robot*: automatically performs an intervention, such as machining bone.
- *Semi-active robot*: performs the intervention under the direct control of the surgeon (e.g., a 'hands on' or 'cooperative control' mode).
- *Passive robot*: does not actively perform any part of the intervention (e.g., positions a tool guide).

There is some debate whether one class of robots may be better than another when considering factors such as safety, user acceptance or regulatory approval. In the latter case, it is likely that the less active a robot is, the more comfortable regulatory agencies will be in granting approval. Regarding safety, although a passive robot may avoid some of the risks inherent with a more active robot, there are still considerable safety issues that must be considered in all cases. For example, when preparing the bone for a knee prosthesis, regardless of whether the bone is automatically machined by an active robot, cooperatively machined by the surgeon and semi-active robot, or machined by the surgeon using a tool guide positioned by a passive robot, it is critical that the cutting be performed at the correct position and orientation. Therefore, each of these robots must provide a safety system to ensure that sensor failures do not cause them to incorrectly position the cutting tool or tool guide. The question of user acceptance has not yet been answered because currently, the difficulty of using medical robots has been a bigger obstacle than whether they are active, semi-active or passive.

2.5   Intraoperative Human–Machine Interfaces

Fundamentally, CIIM systems are intended to work *with* clinicians, not *replace* them in the operating room or interventional suite. Consequently, technology and methods for human – machine communication are crucial components in these systems. This communication is two-way, and successful systems must address techniques both for *providing information* to and for *accepting information and direction* from the clinician.

Visual display is the most common method for providing information to the clinician. Computer displays relating the positions of surgical instruments to cross-sectional medical images or to x-ray projections are ubiquitous in surgical navigation systems (see Section 4.2). The ergonomics of such systems have some serious limitations. Once a procedure has begun, the clinician's attention is necessarily focused on the patient's anatomy, and it is awkward for the clinician to look away from the patient. Consequently, a number of groups have developed systems and devices for superimposing visual information directly on the surgeon's view of the patient. The first of such systems (e.g., [30–32]) were designed to 'inject' registered graphic information into a surgical microscope. Subsequently, several groups have developed variations on this theme for use in other environments (see Fig. 3). Some of these systems may use active elements such as laser pointers (e.g., [33, 34]) to help the surgeon achieve a desired alignment. Other forms of feedback used by CIIM systems include auditory feedback, either in the form of computer-generated speech [35] or simple auditory cues [36], haptic (force) feedback [37–40] or visual/auditory representation of tool–tissue interaction forces (e.g., [41]).

There are many ways for a surgeon to provide information or command direction to a CIIM system. The most common are those used with any computer workstation: typed text and mouse-like pointing devices. Intraoperatively, these devices have many limitations, especially because they are difficult to sterilize and they tie up the clinician's hands. One common, though clearly limited, work-around has been to rely on verbal instructions to technicians operating the equipment. Another has been to rely on computer voice recognition systems (e.g., [46–49]). Still another has been to rely upon sterile touch screen displays or upon the motions of instruments tracked by surgical navigation systems. A few groups have explored video tracking of the clinician's head or eye motions (e.g., [50]).

The motion of surgical robots is frequently commanded through the use of conventional telerobotic 'master' devices, which are essentially powered or unpowered robot manipulators moved by the clinician, or by 'cooperative' control methods in which the robot's motion complies to forces exerted on it by the clinician (see Section 4.3). Other methods, often used in research systems designed for more 'intelligent' assistance to a surgeon, include visual tracking of surgical instruments and target anatomy (e.g., [51, 52]).

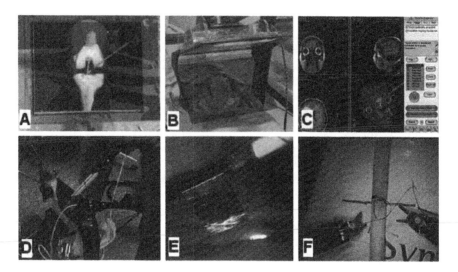

Fɪɢ. 3. Visual information display in CIIM systems. A) CMU image overlay system [42] based on active tracking of surgeon's head, 3D graphics, and semi-transparent mirror; B) JHU image overlay system for simple in-scanner display of scan planes [43, 44]; C) typical display from a surgical navigation system (courtesy, Medtronic Navigation); D) Osaka/Tokyo laser guidance system [33]; E) JHU/Intuitive Surgical overlay of laparoscopic ultrasound onto daVinci surgical robot video monitor [45]; F) Sensory substitution display of surgical force information onto daVinci surgical robot video monitor [41].

2.6 Sensorized Instruments

A number of research groups (e.g., [56–63]) have developed 'sensorized' surgical instruments capable of measuring tool-to-tissue interaction forces and providing these results to surgical workstations. Often, these efforts have relied on graphical interfaces to display force data, whether the instrument was manipulated freehand by the clinician or by a robot. For example, Poulose et al. [58, 59] demonstrated that a force sensing instrument used together with an IBM/JHU LARS robot [51] could significantly reduce both average retraction force and variability of retraction force during Nissen fundoplication. There have also been efforts to incorporate sensed force information into the control of robotic devices (e.g., [64–67]). Several researchers (e.g., [68, 69]) have focused on specialized 'fingers' and display devices for palpation tasks requiring delicate tactile feedback (e.g., for detecting hidden blood vessels or cancerous tissues beneath normal tissues). Yet another use of sensorized instruments is in biomechanical studies to measure organ and tissue mechanical properties to improve surgical simulators (e.g., [70,71]).

There has also been work to integrate non-haptic sensors with surgical instruments. For example, our group at Johns Hopkins is developing instruments that measure

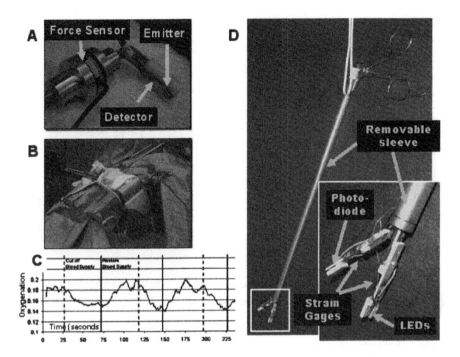

FIG. 4. Sensorized instruments from our laboratory at JHU [53–55]. A) Liver retractor with integrated force sensor and optical sensors for measuring blood oxygenation. B) Retraction of pig liver; C) Sensor readings as blood supply is cut off and restored; D) Laparoscopic instrument with force and oxygenation sensing fingers.

tissue oxygenation as well as force [53–55]. Our plan is to use this information to help surgeons assess tissue viability, avoid ischemic tissue damage during retraction and distinguish tissue types (see Fig. 4).

2.7   Software and Robot Control Architectures

Figure 5 shows the basic control architecture for a robot system There are two periodic loops: a high-frequency servo loop (typically 1 KHz or higher) that controls the individual motors and a low-frequency supervisory loop (typically about 100 Hz) that coordinates the individual motors and may also close a loop around an external sensor, such as a force sensor or imaging system. Although the dynamic equations of a robot include coupling between the axes, the standard practice is to perform the servo control of each motor independently. In fact, some robot systems perform the servo control on a distributed network of embedded microprocessors, where each

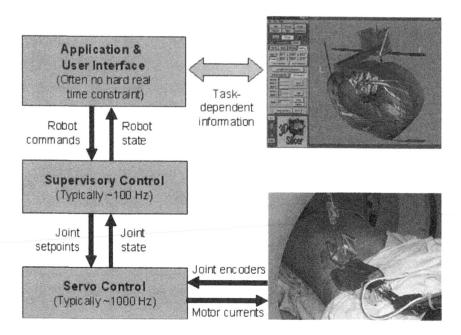

FIG. 5. Architecture of a typical robot system. The robot shown was developed at Johns Hopkins for in-scanner percutaneous needle placement procedures [19,72,73]. The screen interface at the top is typical of the sort of research interface commonly developed for similar procedures, although a different interface was used for the kidney biopsy shown at the bottom.

microprocessor is attached to just one or two motor/sensor pairs. Fortunately, most medical robots move rather slowly (often for safety reasons), so the dynamic coupling between joints can be ignored without affecting control performance.

For a typical path-controlled robot, the supervisory control loop consists of a trajectory planner that breaks down a high-level motion command (such as moving at a specified velocity along a straight line) into a set of intermediate set points that are sent to the servo control loop(s). Because the high-level motion command is often in a Cartesian coordinate system, this process generally includes the invocation of the robot's inverse kinematic equations to transform the Cartesian coordinates into the robot's joint coordinates expected by the servo loop. Although path-controlled robots are common in industrial applications, in the medical field many other supervisory control strategies are often required. One example is a compliant control mode, where robot motion is dictated by the forces and torques applied by the surgeon and measured by a force sensor. This is often implemented as an admittance controller, where the measured forces/torques are multiplied by an admittance gain to produce a desired Cartesian velocity. Typically, the Cartesian velocity is transformed to joint velocities via the robot's inverse Jacobian. Alternatively, the Cartesian velocity can be added to the current Cartesian position to obtain a desired position, which can be transformed to joint positions via the robot's inverse kinematics.

Historically, robot manufacturers have provided an interpreted language for programming the robot, since this allows the end-user (or systems integrator) to quickly develop new applications or modify existing applications in the field, if necessary. An interpreter environment allows fast implement-test-debug cycles, especially since debugging changes can be made during execution without losing any of the program state, such as the robot's position [74]. On the other hand, compiled code runs significantly faster than interpreted code, which is especially important for tasks that require real-time performance, such as closing a loop around an external sensor. For medical robots, regulatory requirements must also be considered – these dictate that a medical device manufacturer must carefully control all software changes (configuration control). This requirement and liability considerations necessitate a system design that prevents inadvertent or unauthorized software modifications, especially in the field. Although it is possible to protect interpreted code from modification (e.g., by encryption of the source code), a compiled language has the advantage of enabling manufacturers to provide end-users with only the executable. For these reasons (efficiency and security), most medical robots are programmed in a compiled language such as C or C++. For development, however, it is still desirable to have an interactive (interpreted) environment. This can be achieved by 'wrapping' the C/C++ code for use with a standard interpreted language, such as TCL or Python.

The development of standard software and control libraries and application frameworks for medical robotics research represents a significant challenge and opportunity. This goal has been a major research focus at Johns Hopkins University. Figure 6 illustrates the software/hardware environment being developed in our laboratory at Johns Hopkins for research on 'intelligent' surgical assistants, which is based on the set of open source software libraries which we are developing (e.g., [75–77]). The development of this sort of infrastructure can be an important 'enabler' in medical robotics research. See also Section 4.6.

2.8 Accuracy Evaluation and Validation

Validation of computer-integrated interventional systems is challenging because the key measure is how well the system performs in an operating room or interventional suite with a real patient. Clearly, for both ethical and regulatory reasons, it is not possible to defer all validation until a system is used with patients. Furthermore, it is often difficult to quantify intraoperative performance because there are limited opportunities for accurate post-operative assessment. For example, even though

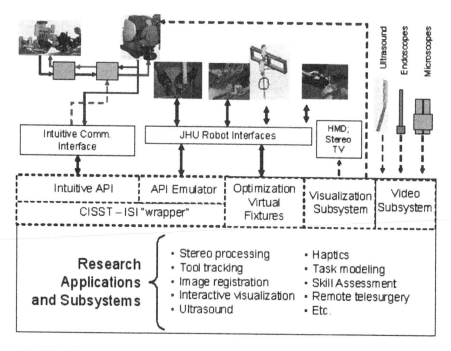

FIG. 6. Modular system environment for robotic surgical assistance research at Johns Hopkins University (http://www.cisst.org/cisst).

CT scans are accurate, they may not provide sufficient contrast for measuring the post-operative result and they expose the patient to additional radiation. For these reasons, most computer-integrated interventional systems are validated using *phantoms*, which are objects that are designed to mimic (often very crudely) the relevant features of the patient.

One of the key drivers of Surgical CAD/CAM is the higher level of accuracy that can be achieved using some combination of computers, sensors and robots. Therefore, it is critical to be able to evaluate the overall accuracy of such a system. One common technique is to create a phantom with number of objects whose locations are accurately known, either by precise manufacturing or measurement. If the system uses fiducial-based registration, the objects in the phantom should correspond to fiducials. Furthermore, the phantom should contain extra fiducials (not used for registration) or other known features that can be used as targets. If the system uses an anatomic registration, it may still be useful to place a number of fiducials in the phantom so that they can provide a reasonably accurate estimate of the 'ground truth' registration.

The basic technique is to image the phantom, perform the registration and then locate the target features. Maurer [78] defined the following types of error:

- *Fiducial Localization Error (FLE)*: the error in locating a fiducial in a particular coordinate system (i.e., imaging system or surgical CAD/CAM system).
- *Fiducial Registration Error (FRE)*: the root-mean-square (RMS) residual error at the registration fiducials, i.e.,

$$FRE = \sqrt{\frac{1}{N} \sum_{k=1}^{N} \left\| \vec{\mathbf{b}}_k - \mathbf{T} \cdot \vec{\mathbf{a}}_k \right\|^2}$$

where $\mathbf{T}$ is the registration transform and $(\vec{\mathbf{a}}_k, \vec{\mathbf{b}}_k)$ are matched pairs of homologous fiducials ($k = 1, \ldots, N$).

- *Target Registration Error (TRE)*: the error in locating a feature or fiducial that was not used for the registration; if multiple targets are available, the mean error is often reported as the TRE.

For a robot system, one method for measuring TRE is to locate the targets in the image, transform them to the robot coordinate system (using the registration) and then command the robot to position its instrument at the computed target location. The TRE is given by the difference between the robot's position and the actual position of the target. It may not be practical or convenient to measure this position difference; however, a common strategy is to manually position the robot at the physical target and then compute the TRE as the difference between the computed position (based on the registration) and the robot's actual position. Essentially, this method uses the robot itself to measure the TRE.

2.9   Risk Analysis and Regulatory Compliance

The medical device industry is a heavily regulated industry. In the United States, medical devices must be cleared for market by the Food and Drug Administration (FDA). There are two paths to market: one via the 510(K) premarket notification process and one via the Pre-Market Approval (PMA) process. A manufacturer can obtain a 510(K) clearance if the new device is 'substantially equivalent' to an existing device that is already on the market. Otherwise, the PMA application is required. Surprisingly, several medical robots obtained clearance via the 510(K) path, including Aesop (Computer Motion, Inc.), Neuromate (Innovative Medical Machines International and subsequently Integrated Surgical Systems) and daVinci (Intuitive Surgical). In contrast, the earlier ROBODOC System (Integrated Surgical Systems and subsequently Curexo Medical) started down the PMA path, although the company was later able to switch to the 510(K) process. As of March 2008, the ROBODOC 510(K) application is still pending.

In addition to the need for 510(K) or PMA approval, medical device companies must comply with the Quality System Regulations (QSR) and are periodically audited by the FDA to verify compliance. Initially, the FDA required companies to adhere to Good Manufacturing Practices (GMP), which regulated just the manufacturing phase. For simple devices, this worked well because device failures were primarily due to manufacturing flaws. As devices became more complex, especially with the integration of computers and software, FDA discovered that a large number of device failures were due to design flaws, rather than manufacturing flaws. The infamous Therac-25 accident, where six patients received massive overdoses of radiation from a computer-controlled medical linear accelerator, is a well-known example [79]. As a result, FDA QSR began to regulate the design phase as well.

In the European market, all products (medical or otherwise) require CE marking. Furthermore, the design and manufacturing processes must comply with ISO 9001 and 9002, respectively (often these are grouped together under the umbrella term ISO 9000). The CE marking and ISO 9000 certification are handled by a number of *notified bodies*, which are independent (non-governmental) entities.

To comply with ISO 9000 and/or FDA QSR, medical device companies must define their development and manufacturing processes and then produce documents (quality system records) that demonstrate adherence to these processes. Although ISO 9000 and FDA QSR are similar, they are not identical, which requires most medical device companies to comply with both of them.

It should be noted that obtaining FDA approval and CE marking and complying with FDA QSR and ISO 9000 is still not enough to guarantee commercial success. Obviously, it is necessary for the device to be marketable (i.e., to provide a favourable cost–benefit ratio). It is perhaps less obvious, however, that the device must also be accepted by the third-party payers in the health-care system. In the U.S., this consists of Medicare and the health insurance companies. These entities must agree to reimburse for procedures performed with the new technology in order for that technology to proliferate in the marketplace.

Risk (or Hazard) Analysis is one of the key elements of a medical device development process and is often a focal point for audits by FDA or notified bodies. A Failure Modes Effects Analysis (FMEA) or Failure Modes Effects and Criticality Analysis (FMECA) is the most common method [80]. These are 'bottom up' analyses, where potential component failures are identified and traced to determine their effect on the system. Methods of control are devised to mitigate the hazards associated with these failures. The information is generally presented in a tabular format. The FMECA adds the 'criticality' assessment, which consists of three numerical parameters: the severity (S), occurrence (O) and detectability (D) of the failure. A risk priority number

(RPN) is computed from the product of these parameters; this determines whether additional methods of control are required. The FMEA/FMECA is a proactive analysis that should begin early in the design phase and evolve as hazards are identified and methods of control are developed.

3. Surgical CAD/CAM

3.1 Example: Robotically Assisted Joint Reconstruction

The relative rigidity of bone and the excellent contrast available in X-ray and CT images make orthopaedic procedures – especially joint replacement surgery – natural applications for medical robots and about 20% of all medical robots surveyed in 2005 were intended for such applications [81]. The authors of this chapter were co-developers of one of the first robotic systems for orthopaedics (ROBODOC® [82,83]), so it is natural for us to use it as an example in discussing surgical CAD/CAM applications. Earlier research using a robot for total knee replacement surgery was performed at the University of Washington [84] and subsequently, a number of other groups also developed systems for similar applications (e.g., [85–89]).

ROBODOC® (ROBODOC, a Curexo Technology Company, formerly Integrated Surgical Systems, Inc., Sacramento, CA) was initially developed for Total Hip Replacement (THR) surgery [90, 91] and was later applied to Total Knee Replacement (TKR) [92]. THR surgery involves the preparation of an elongated cavity in the femur (thigh bone) and a rounded cavity in the acetabulum (hip socket) to accommodate the two components of a hip prosthesis: the femoral stem (Fig. 7, Right) and acetabular cup. Accurate placement of components relative to the patient's bones is very important for achieving a good result. Furthermore, with cementless implants, the bone must be shaped to achieve a close fit between the implant and the bone in order to encourage the bone to grow into a porous coating on the implant.

For conventional THR surgery, preoperative planning is performed by overlaying templates (outlines) and by making measurements on two-dimensional X-rays. Templates are available at different magnification factors so that errors due to X-ray magnification can be minimized. Usually, planning is limited to identifying an approximate range of implant sizes and the approximate desired implant position relative to the bone. During surgery, the bone is prepared using hand-held reamers (drills) and broaches to create the desired cavities. Proper execution relies on a significant amount of experience and 'surgical feel', especially when the femoral cavity is prepared. In this case, the surgeon typically begins with the reamer and broach corresponding to the smallest planned implant size. If the cavity feels 'loose'

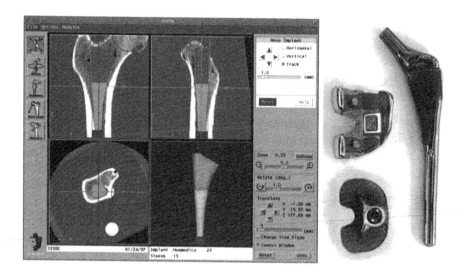

FIG. 7. (Left) Typical screen view from ORTHODOC® CT-based planning system for ROBODOC® orthopaedic robot (Integrated Surgical Systems, Sacramento, California); (Right) Typical implant components for cementless hip and knee reconstruction surgery.

(i.e., insufficient contact with hard cortical bone), the surgeon switches to the next larger size until he/she feels that there is sufficient, but not excessive, cortical contact. If the surgeon chooses a prosthesis that is too large, the femur can fracture either during cavity preparation or during prosthesis insertion. This is one of the most common intraoperative complications associated with THR. Similarly, although the surgeon can plan any desired prosthesis position, the actual position is determined mostly by anatomical constraints because the hand-held instruments tend to follow the path of least resistance.

Laboratory tests [93] showed that the conventional method for cavity preparation is inherently inaccurate. The cavities produced were extremely irregular, with large gaps between implant and bone. Furthermore, accurate alignment of the cavity relative to bone was extremely uncertain, since the interior surface of the bone could deflect the path of the broach. These considerations led our surgeon colleagues (Dr. Paul and Dr. Bargar) to propose the use of a robot to prepare the implant cavity. Expected benefits included adequate and uniform bone in growth, uniform stress transfer, reduced stress shielding, less thigh pain and the elimination of femoral fractures as an intraoperative complication.

The ROBODOC procedure for THR (and TKR) consists of two phases: a preoperative planning phase (ORTHODOC®) and an intraoperative (ROBODOC) phase. The input to ORTHODOC consists of a CT scan of the patient's anatomy, the prosthesis geometry that is supplied by the manufacturers and clinical decisions made by the surgeon. The surgeon plans the procedure by selecting a prosthesis from the database and positioning it in the CT image. ORTHODOC displays three orthographic views (i.e., orthogonal slices) of the data as well as a 3-D model (see Fig. 7). Each joint of the five-axis surgical robot (Fig. 8-A,B) contains two optical encoders for redundant position feedback. The system includes a wrist-mounted six-axis force sensor that monitors the forces applied at the tool. This force information makes it possible to implement functionality such as manual guidance, tactile search, safety checking and an adaptive cutter feed rate. ROBODOC executes the preoperative plan by machining the specified prosthesis cavity in the femur. This requires the bone to be rigidly attached to the robot. A bone motion monitor (BMM) is used as a safety sensor to detect motion of the bone relative to the robot. In addition, accurate cavity placement requires a registration between the patient's anatomy in the preoperative plan (i.e.,

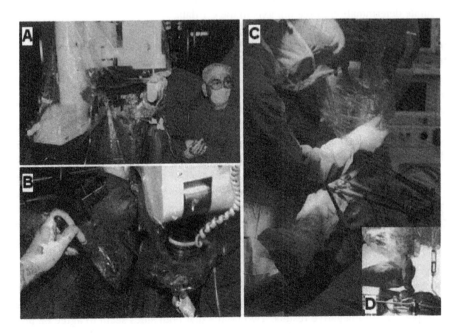

FIG. 8. Clinically applied robots for orthopaedic surgery. A, B) the Robodoc® system for cementless total hip and knee replacement surgery machines bone to match a surgeon-selected implant shape, according to a presurgical plan based on patient's CT images [83, 94]; C, D), the Acrobot system [85] employs cooperative hand guiding with 'active constraints' derived from the implant shape for total knee replacement surgery.

the bones in the CT scan) and the anatomy of the actual patient. The preoperative plan is specified in image (CT) coordinates, whereas intraoperative localization of the patient can be obtained in robot coordinates, so registration implies finding the transformation between image and robot coordinates.

Initially, ROBODOC used a 'pin-based' registration method, which required the implantation of titanium bone screws (pins) in the femur prior to the CT scan. Registration was accomplished by defining at least three reference points on the pins and then identifying them in both the CT and robot coordinate systems. Because the pins are titanium, the ORTHODOC software could easily locate them in the CT data using image processing techniques. The robot system identified the physical pins via a tactile search, using feedback from its wrist-mounted force sensor [83]. ROBODOC initially used three registration pins, with the centres of the pin heads serving as the three reference points. Shortly afterwards, it transitioned to a two-pin method, where the third reference point was obtained by creating a 'virtual pin' based on the centre and axis of the distal pin. In this case, a longer distal pin was required to enable accurate determination of the pin axis in the CT data.

Although pin-based registration is reliable, it involves an extra (minor) surgery to implant the pins prior to the CT scan and is also the source of postoperative knee pain for many patients. This motivated the development of a 'pinless' system [95], which uses anatomical features instead of metal pins as fiducials. Registration is performed using a method similar to the iterated closest point method outlined in Section 2.2, using bone surface point positions measured by a small digitizing arm.

Once cutting has commenced, ROBODOC provides a visual display of its progress on the computer monitor. As the robot mills the cavity, the monitor displays the CT data overlaid with a model of the prosthesis cavity. The completed portion of the cavity is displayed in one colour, while the remaining portion is displayed in another. This is similar to the visualization provided by most navigation systems. During cutting, the control software continuously monitors the force sensor and adjusts the cutter feed rate based on the sensed force and on parameters specific to the prosthesis design and cutting tool [96]. This enables the robot to adapt to the patient's anatomy by slowing down in regions of hard cortical bone and speeding up in other regions.

As of March 2008, ROBODOC has been installed in about 50 hospitals around the world and has performed over 20,000 hip and knee surgeries. Use of this system became controversial, especially in Germany, with surgeons and patients reporting both positive and negative results. Two points that both sides seem to agree on are: 1) the robot procedure requires a longer surgery time and results in higher surgical costs, compared with the conventional technique and 2) the robot can execute the preoperative plan more accurately than the conventional technique. However, there is no consensus on whether the improved accuracy provided by the robot system provides a clinical benefit to the patient.

3.2   Example: Needle Placement

Placement of needles or similar devices[2] is one of the most basic interventional CAD/CAM applications, though there are numerous challenges depending on the target organ and the operating environment. The problem can be simply stated as placing the tip of the needle at a location specified on an image, typically through an entry point that is also specified on the image. Both robots and navigation systems have been used to assist with this task. In some cases, the interventional device (robot or navigation system) is used to position a cannula or instrument guide through which the needle is manually advanced. Accurate placement of needles in the brain was one of the first uses of robots in interventional medicine (e.g., [97, 98] [99, 100]), and since then these techniques have been extended to many parts of the body, including prostate, liver, spine, etc. Furthermore, percutaneous needle placement is a natural application for surgical navigation and 'image overlay' techniques such as those illustrated in Fig. 3. There is an extensive literature on robotic and non-robotic systems for needle placement. This section will touch briefly on a few common themes.

When performed using CAD/CAM techniques, the entry and target positions can be identified on preoperative images, intraoperative images or some combination of the two. In all cases, it is necessary to register the image space to the interventional device (e.g., robot or tracked instrument). When using intraoperative images, this registration can be obtained by placing a calibration object on the robot or patient. The transformation between the calibration object and device coordinate system is known by design, and the transformation between the calibration object and the image coordinate system is computed by locating features of the calibration object in the image, often via image-processing techniques. Recent examples from the work of our own group at Johns Hopkins include [101, 102], although these techniques are widely practiced by many groups.

Another potential issue for CAD/CAM needle placement is target motion. This is especially challenging for soft-tissue organs such as liver, kidneys or prostate, as well as for anatomical targets such as the lungs or spine, which can be affected by respiratory motion or by heart beats. For this reason, many groups have emphasized placement of needles under direct feedback from imaging modalities such as X-ray fluoroscopy, CT, MRI or ultrasound. Whether or not direct image feedback is available, it is often important to compensate for motion and/or to register preoperative images with (possibly deformed) intraoperative anatomy or images.

In many cases, needle placement under direct (intraoperative) image guidance is difficult due to patient access issues. This is especially true when the image modality

---

[2] For convenience, we use the term 'needle placement', but the problem is generic to the placement of any 'needle-like' instrument, including probes, drills, radiation beams, etc.

is a closed-bore MR scanner, where the patient is placed inside a long cylindrical tube that has a diameter that is not much larger than the patient. Here, the only option for performing needle placement (besides catheter-based methods) is to use a robot that is small enough to fit inside the MRI scanner (e.g., [27, 103]). The design of MR-compatible robotic devices poses significant challenges due to materials and component limitations associated with the high magnetic fields and radiofrequency sensing associated with MR imaging (e.g., [104]). Even for a robot intended for use with CT or X-ray fluoroscopy, it is generally desirable for the robot's end effector to be as radiolucent as possible in order to reduce interference with the images used for guidance and targeting.

4. Surgical Assistance

4.1 Basic Concepts

Interventional procedures – especially those that we think of as 'surgery' – can be highly interactive processes, and many interventional decisions are made in the operating room and executed immediately. The goal of computer-based interventional systems, including medical robots, is not to replace the surgeon or interventionist[3] with a machine so much as to provide the surgeon with versatile tools that augment his or her ability to treat patients. Currently, there are three main sub-classes of Assistant Systems, although the distinctions between them are by no means hard-and-fast.

The first class, *Intraoperative Information Support Systems*, simply provides information to the surgeon, who uses his or her manual dexterity to manipulate the surgical instruments in performing the intervention. An extremely important sub-class of these systems (discussed in Section 4.2) is *Surgical Navigation Systems*, which relate surgical instrument positions to medical images and patient anatomy.[4]

The second class, *Surgeon Extender Robots*, are operated directly by the surgeon and augment or supplement the surgeon's ability to manipulate surgical instruments. Potentially, these systems can give even average surgeons super-human capabilities such as elimination of hand tremor or ability to perform dexterous operations inside the patient's body. The potential clinical advantages associated with these systems include a) the ability to treat otherwise untreatable conditions; b) reduced invasiveness and patient morbidity; c) improved safety and reduced complication rates; and d) reduced

---

[3] For simplicity of discussion, we will use the word 'surgeon' throughout the balance of this section, rather than the more inclusive (but awkward) 'interventionist'.

[4] Interestingly, Surgical Navigation systems can also be thought of as Surgical CAD/CAM, since they provide the capability to couple presurgical image-based planning with intraoperative execution.

surgeon fatigue. A special sub-class of surgeon extender robots is *Remote Telesurgery Systems*, which permit the surgeon to operate on patients at distances ranging from a few hundred meters to several thousand kilometers.

A third class, *Auxiliary Surgical Supports*, generally works side by side with the surgeon and performs such functions as endoscope holding, tissue retraction or limb positioning. These systems typically provide one or more direct control interfaces such as joysticks, head trackers, or voice control. However, there have been some efforts to make these systems 'smarter' so as to require less of the surgeon's attention during use, for example, by using computer vision to keep the endoscope aimed at an anatomic target or to track a surgical instrument. Although these systems may offer some of the same advantages as surgeon extenders (e.g., reduced tissue damage due to more delicate retraction), their main justification is improved operative efficiency and reduced need of operating room staff.

4.2 Surgical Navigation Systems as Information Assistants

Surgical navigation systems track the positions of surgical instruments and other objects in the operating room and display this information graphically, usually relative to registered images of the patient. Although first developed for neurosurgery (e.g., [8, 105]) they have also been widely adapted to ENT surgery (e.g., [11, 106]), orthopaedic surgery (e.g., [9, 10]), craniofacial surgery (e.g., [107, 108]) and other applications placing a high value on precise localization and integration of information from medical imaging systems. There are currently many commercially available systems, and surgical navigation has rather larger acceptance in the interventional systems market than does any form of robotic assistance.

As shown in Fig. 9, a typical surgical navigation system consists of a navigational tracking device capable of determining the position and orientation of 'rigid bodies' attached to surgical instruments and to the patient's anatomy, together with a computer workstation and display. After a registration step is performed, the workstation is able to compute and display the position of instruments relative to patient images.

4.3 Surgeon Extenders

Telesurgical robots are the most widely deployed form of surgeon extender system and have been extensively employed for cardiac, prostate, and other minimally invasive laparoscopic procedures. Examples include numerous (dozens) research systems (e.g., [20, 25, 109–114]), as well as commercially deployed systems such the daVinci

Workstation with Display   Tracking device

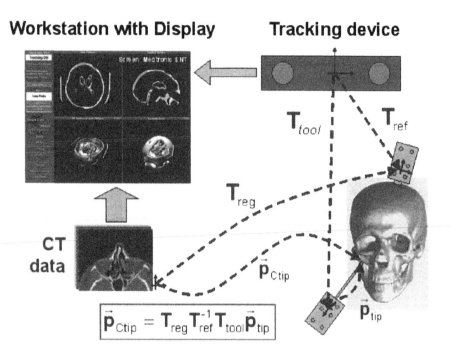

$$\vec{P}_{Ctip} = T_{reg} T_{ref}^{-1} T_{tool} \vec{P}_{tip}$$

FIG. 9. A typical surgical navigation system, showing key coordinate transformations. After registration, the system computes $\vec{P}_{Ctip}$, the position in image coordinates corresponding to the current position of the pointer tip, and uses this information to update a display.

[115] (Intuitive Surgical, Sunnyvale, Ca.) and Zeus [116] (formerly marketed by Computer Motion, Goleta, Ca.).

The architecture of a typical system (here, the daVinci) is shown in Fig. 10. The system consists of a patient-side 'slave' robot and a master control console. The slave robot has three or four robotic arms which manipulate a stereo endoscope and dexterous surgical instruments such as scissors and needle holders. The surgeon sits at the master control station and grasps handles attached to two dexterous 'master' manipulator arms, which are capable of exerting limited amounts of force feedback to the surgeon. The surgeon's hand motions are sensed by the master manipulators and the motions are mimicked by the slave manipulators. A variety of control modes may be selected by means of foot pedals on the master console and used for such purposes as determining which slave arms are associated with the hand controllers. Stereo video is transmitted from the endoscope to a pair of high-quality video monitors in the master control station, thus providing high-fidelity stereo visualization of the surgical site. The display and master manipulators are arranged so that it appears to

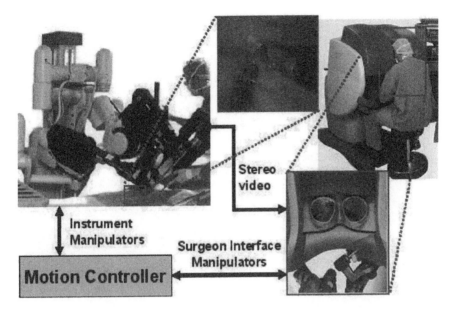

FIG. 10. Architecture of a typical telesurgical system. Photos: Intuitive Surgical Systems.

the surgeon that the surgical instruments (inside the patient) are in the same position as his or her hands inside the master control console. Other telesurgical systems employ the same basic architecture, although there are many differences in implementation. For example, many systems (e.g., [26, 116]) use more conventional 'stereo TV set' displays that use polarizing glasses or LCD shutter glasses to multiplex left- and right-eye images. Some surgeons find this arrangement more comfortable for long-duration procedures, although much of the 'immersive' feel of the daVinci is lost. Similarly, research systems incorporate many different mechanical designs for the patient-side 'slave' robots.

A primary advantage promised by telesurgical robots for minimally invasive surgery is their ability to permit the surgeon to perform dexterous manipulation of instruments and tissues inside the patient's body. A major theme in current research has been the development of highly dexterous, miniaturized robotic end-effectors suitable for this purpose. Some examples are shown in Fig. 11. Many systems (e.g., [20, 25, 115, 121, 122]) have used cable actuated tools. One drawback of this approach is that it becomes increasingly difficult to provide high strength and dexterity as the mechanisms get smaller and smaller. This has led various groups to investigate alternatives. For example, several groups have explored micro-hydraulic systems (e.g., [123, 124]). At Johns Hopkins, we have explored another approach, illustrated in

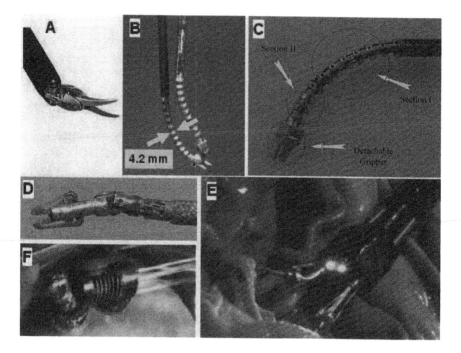

Fɪɢ. 11. Dexterity and mobility inside the patient's body. A) daVinci dexterous wrist with typical surgical instrument (courtesy: Intuitive Surgical); B, C) 4.2 mm diameter JHU/Columbia University 'Snake' manipulator [128, 129]; D) 5 dof, 3 mm diameter micro-catheter robot [112, 121]; E) dexterous robot for endogastric surgery [122]; F) mobile 'Heart Lander' robot for crawling across the heart [127].

Fig. 11 B-C, using parallel super-elastic spines to produce 'snake'-like end-effectors [123]. Although most current surgical robots employ manipulator arms to position tools within the patient's body, with wrist-like mechanisms to provide distal dexterity, there has been some work on systems with a greater degree of autonomous motion capability (e.g., [125–127]).

Although teleoperation has many advantages, especially for high-dexterity robotic manipulation inside the patient's body, it also has some drawbacks. The amount of equipment required is large, since both 'master' and 'slave' manipulators are needed. The surgeon is frequently somewhat removed from the patient, since he or she is sitting at a master control station and may have a reduced overall awareness of the surgical situation.

Consequently, several groups, including our own, have developed an alternative approach based on 'hands-on' admittance control, in which the robot moves in response to forces exerted by the surgeon directly on the robot's end effector or on a handle attached to the robot. Our early experiences with Robodoc® [83] and other surgical robots (e.g., [51, 130]) showed that surgeons find this form of control to be very convenient and natural for surgical tasks. Two notable uses of cooperative control are the Imperial College Acrobot™ orthopaedic system [85] (Fig. 8 C–D) and the Johns Hopkins 'Steady Hand' microsurgery system [117] (Fig. 12). Although cooperative control is usually limited to precise positioning tasks, it can also provide force scaling via the use of two force sensors: one to sense the surgeon's input and another to measure tool-to-tissue interaction forces and then move the robot in response to a scaled difference between these forces (e.g., [66, 131]).

Other groups have developed completely free-hand instruments that sense and actively cancel physiological tremor (e.g., [132, 133]). The main advantage of this approach is that it requires the least change in normal operating room procedure. The surgeon uses the tremor-reducing tool just as he or she would use any other instrument. The challenges are instrument ergonomics (mostly size and weight) and precise motion performance, which is still not as good as that of fully robotic devices.

One problem commonly encountered in all forms of medical robotics is the difficulty of maintaining a desired relationship between an instrument held by the robot and moving patient anatomy. Broadly speaking, there are two approaches for solving this problem. The first approach (e.g., [52, 134]) is to sense the relative motion – most

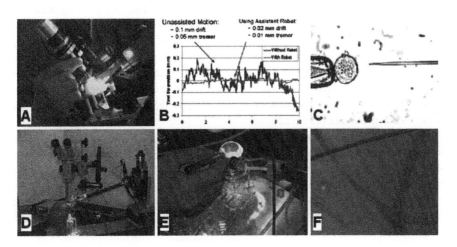

Fig. 12. JHU 'Steady Hand' cooperative manipulation systems for microsurgery. A) First-generation system [117], which is here used to demonstrate fenestration of the stapes bone for an otology application [67]; B) Comparative motion tremor with freehand instrument manipulation and steady-hand robot manipulation [118]; C) Steady-hand micro-injections into mouse embryos [119]; D) Newer generation steady-hand robot for eye surgery [120]; E–F) Evaluation on chick embryos.

commonly with computer vision or some other form of imaging device – and then move the robot. The second approach (e.g., [83, 86, 127, 135, 136]) is to attach the robot's base firmly to the patient's anatomy, so that it rides with the patient. This approach is especially common in orthopaedics, but may fruitfully be applied in other areas such as ENT, neurosurgery, cardiac surgery or ophthalmology, where a good attachment point is available.

4.4   Auxiliary Surgeon Supports

Although attention is often focused on robotic systems that directly extend the surgeon's ability to manipulate surgical instruments, many of the most successful robotic applications in surgery have focused on auxiliary tasks such as patient positioning (e.g., [137]), surgical instrument delivery [138, 139] and laparoscopic camera positioning (e.g., [51,52,140]). In fact, the AESOP® laparoscopic camera surgery system [46,141] (formerly distributed by Computer Motion, Goleta, Ca.) was one of the first widely deployed surgical robots.

4.5   Remote Telesurgery and Telementoring

The possibility for using master-slave telesurgery systems to perform procedures in which the surgeon and patient are separated by very long distances has long been recognized [142, 143]. Commonly considered applications include space exploration, military combat care and provision of care in sparsely populated areas. A number of research groups have developed experimental systems over the years (e.g., [20, 144–149]). A major milestone was achieved by Marescaux et al. in 2001, with successful performance of a trans-Atlantic laparoscopic cholecystectomy [150]. Subsequent work has included efforts by Anvari et al. to develop a practical system for deployment in Canada [151,152].

There has also been significant interest in using telesurgical technology to provide remote (or on-site) mentoring, in which an expert surgeon advises a less experienced surgeon in carrying out a procedure (e.g., [153–155]). Although in some ways similar to more 'conventional' telesurgery, this form of telementoring can introduce some additional challenges. In particular, protocols may be needed to enable the expert and 'trainee' surgeon to trade-off control of a surgical robot or otherwise to work cooperatively during completion of the case.

4.6   Toward 'Intelligent' Surgical Assistance

Although one goal of both teleoperation and hands-on control in a surgeon extender system is to enable the surgeon to directly control the motion of the robot, the fact that

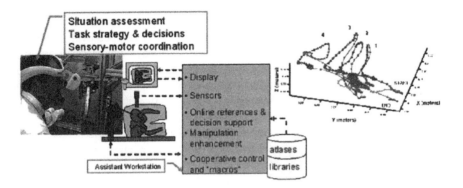

FIG. 13. (Left) Functional Architecture of a typical 'intelligent' surgical assistant; (Right) segmented trace of daVinci hand motions during a suturing procedure [161]

a computer is meditating between the surgeon's command input and the robot's actual motion can create many more possibilities. The simplest is a safety barrier or 'no fly zone', in which the robot's tool is constrained from entering certain portions of its workspace. More sophisticated versions include virtual springs, dampers or complex kinematic constraints that help a surgeon align a tool, maintain a desired force or perform similar tasks. This concept has many names, of which 'virtual fixtures' seems to be the most popular (e.g., [156–160]). The Acrobot system shown in Fig. 8 C–D represents a successful clinical application using virtual barriers to limit the motion of the cutting tool.

A number of groups (e.g., [162–165]) are exploring extensions of the virtual fixtures concept to active cooperative control, in which the surgeon and robot share or trade-off control of the robot during a surgical task or sub-task. As the ability of computers to model and 'follow along' surgical tasks (e.g., [139, 161]) improves, these modes will become more and more important in surgical-assistant applications. Figure 13 (Left) shows the functional architecture of a typical 'surgical assistant' workstation being developed at Johns Hopkins University. Figure 13 (Right) illustrates initial efforts to develop automatic motion segmentation tools to distinguish the different steps in a suturing procedure.

5. Summary and Conclusion

Medical Robotics and computer-integrated interventional medicine are still relatively 'young' fields. Nevertheless, they have grown remarkably, especially in the past 5–8 years, as clinical systems have been deployed and as more researchers enter them.

This short chapter has only provided a brief introduction to some of the main areas of research and practice, and our treatment has necessarily skipped over important research and groups working in the field. To those who may have been left out, we extend our sincere apologies and hope that readers of this chapter will be motivated to pursue further reading, perhaps starting with books such as [166, 167], recent journal special issues such as [114, 168] or any of the many conference proceedings in the field.

By coupling information to action in ways that were not possible before, these systems have the potential to fundamentally change the practice of interventional medicine. Enough progress has been made in all of the architectural elements shown in Fig. 2 so that clinically useful systems can indeed be deployed. However, further advances are still needed across the board in the modelling and analysis required for medical robotic applications, for the interface technologies required to relate the 'data world' to the physical world of patients and clinicians and to the system science that makes it possible to put everything together safely, robustly and efficiently. It is our belief that this research is best done in interdisciplinary teams motivated by important applications. Our experience has been that building a strong researcher-surgeon industry team is one of the most challenging, but also one of the most rewarding aspects of medical robotics and CIIM research. The only greater satisfaction is the knowledge that the results of such teamwork can have a very direct impact on patients' health. This is a challenging area, but it is worth it.

REFERENCES

[1] Taylor R. H., Lavallee S., Burdea G., and Mosges R., 1996. Computer-Integrated Surgery, Cambridge, Mass.: MIT Press.

[2] Taylor R. H., September 2006. A perspective on medical robotics. *IEEE Proceedings*, vol. **94**, pp. 1652–1664.

[3] Taylor R. H., and Joskowicz L., 2003. Computer-integrated surgery and medical robotics. In Kutz M., Ed., *Standard Handbook of Biomedical Engineering and Design*, McGraw Hill, pp. 29.23–29.45.

[4] Taylor R. H., and Stoianovici D., October 2003. Medical robotics in computer-integrated surgery, *IEEE Transactions on Robotics and Automation*, vol. **19**, pp. 765–781.

[5] Maintz J. B., and Viergever M. A., 1998. A survey of medical image registration, *Medical Image Analysis*, vol. **2**, pp. 1–37.

[6] Lavallee S., 1996. Registration for computer-integrated surgery: methodology, state of the art, in Taylor R. H., Lavallee S., Burdea G., and Mosges R., Eds. *Computer-Integrated Surgery*, Cambridge, Mass.: MIT Press, pp. 77–98.

[7] Besl P. J., and McKay N. D., 1992. A method for registration of 3-D shapes, *IEEE Transactions on Pattern Analysis and Machine Intelligence*, vol. **14**, pp. 239–256.

[8] Maciunas R. J., 1993. *Interactive Image-Guided Neurosurgery*: American Association of Neurological Surgeons.

[9] DiGioia A., Jaramaz B., Picard F., and Nolte L. P., 2004. *Computer and Robotic Assisted Knee and Hip Surgery*, Oxford Press.

[10] Stiehl J. B., Konerman W. H., and Haaker R. G., 2003. *Navigation and Robotics in Total Joint and Spine Surgery*. Berlin: Springer.

[11] Metson R., Gliklich R. E., and Cosenza M., 1998. A comparison of image guidance systems for sinus surgery, *The Laryngoscope*, vol. **108**, pp. 1164–1170.

[12] Chassat F. L., S., 1998. Experimental protocol for accuracy evaluation of 6-d localizers for computer-integrated surgery: application to four optical localizers. In *Medical Image Computing and Computer-Aided Interventions (MICCAI)*, Cambridge, Mass., pp. 277–284.

[13] Li Q., Zamorano L., Jiang Z., Gong J., Pandya A., Perez R., and Diaz F., 1999. Effect of optical digitizer selection on the application accuracy of a surgical localization system. *Comput Aided Surg.*, vol. **4**, pp. 314–321.

[14] Troccaz J., Peshkin M., and Davies B. L., 1997. The use of localizers, robots, and synergistic devices in CAS. In *Proc. First Joint Conference of CVRMed and MRCAS*, Grenoble, France, pp. 727–729.

[15] Peshkin M. A., Colgate J. E., Wannasuphoprasit W., Moore C. A., Gillespie R. B., and Akella P., 2001. Cobot architecture, *IEEE Transactions on Robotics and Automation*, vol. **17**, pp. 377–390.

[16] Taylor, R. H., Paul H. A., Cutting C. B., Mittelstadt B., Hanson W., Kazandides P., Musits B., Kim Y.-Y., Kalvin A., Haddad B., Khoramabadi D., and Larose D., 1992. Augmentation of Human Precision in computer-integrated surgery. *Innovation et Technologie en Biologie et Medicine*, vol. **13**, pp. 450–459.

[17] Davies B. L., Hibberd R. D., Timoney A. G., and Wickham J. E. A., 1996. A clinically applied robot for prostatectomies. In *Computer Integrated Surgery: Technology and Clinical Applications*: MIT Press, pp. 593–601.

[18] Taylor R. H., Funda J., Eldridge B., Gruben K., LaRose D., Gomory S., and Talamini M. D., Mark, 1996. A telerobotic assistant for laparoscopic surgery. In Taylor R., Lavallee S., Burdea G., and Moesges R., Eds., *Computer-Integrated Surgery*, MIT Press, pp. 581–592.

[19] Stoianovici D., Whitcomb L., Anderson J., Taylor R., and Kavoussi L., 1998. A modular surgical robotic system for image-guided percutaneous procedures. In *Medical Image Computing and Computer-Assisted Interventions (MICCAI-98)*, Cambridge, Mass. pp. 404–410.

[20] Mitsuishi M., Watanabe T., Nakanishi H., Hori T., Watanabe H., and Kramer B., 1995. A telemicrosurgery system with colocated view and operation points and rotational-force-feedback-free master manipulator. In *Proc. 2nd Int. Symp. on Medical Robotics and Computer Assisted Surgery*, Baltimore, Md., pp. 111–118.

[21] DiMaio S., Fischer G., Haker S., Hata N., Iordachita I., Tempany C., and Fichtinger G., 2006. Design of an prostate needle placement robot in MRI scanner. In *IEEE International Conference on Biomedical Robotics*, Pisa, Italy.

[22] Krieger A., Fichtinger G., Metzger G., Atalar E., and Whitcomb L. L., 2006. A hybrid method for 6-DOF tracking of MRI-compatible robotic interventional devices. In *IEEE International Conference on Robotics and Automation*, Orlando, Florida.

[23] Chinzei K., Gassert R., and Burdet E., 2006. Workshop on MRI/fMRI compatible robot technology – a critical tool for neuroscience and image guided intervention. In *IEEE Int. Conference on Robotics and Automation*, Orlando, Florida.

[24] Hempel E., Fischer H., Gumb L., Hohn T., Krause H., Voges U., Breitwieser H., Gutmann B., Durke J., Bock M., and Melzer A., 2003. An MRI-compatible surgical robot for precise radiological interventions, *Comput Aided Surg.*, vol. **8**, pp. 180–191.

[25] Harada K., Tsubouchi K., Fujie M. G., and Chiba T., 2005. Micro manipulators for intrauterine fetal surgery in an open MRI. In *IEEE International Conference on Robotics and Automation (ICRA)*, Barcelona, Spain, pp. 504–509.

[26] Louw D. F., Fielding T., McBeth P. B., Gregoris D., Newhook P., and Sutherland G. R., 2004. Surgical robotics: a review and neurosurgical prototype development. *Neurosurgery*, vol. **54**, pp. 525–537.

[27] Stoianovici D., Patriciu A., Doru Petrisor, Dumitru Mazilu, Muntener M., and Kavoussi L., 2006. MRI-guided robot for prostate interventions. In *Society for Minimally Invasive Therapy (SMIT) 18th Annual Converence*, Pebble Beach.

[28] Davies B., 1996. A discussion of safety issues for medical robots. In Taylor R., Lavallee S., Burdea G., and Moesges R., Eds., *Computer-Integrated Surgery*, Cambridge, Mass.: MIT Press, pp. 287–296.

[29] Picard F., Moody J., and DiGioia A., 2004. Clinical classification of CAOS systems in *Computer and Robotic Assisted Knee and Hip Surgery*, Oxford University Press, pp. 43–48.

[30] Frets E. M., Strobe J. W., Hatch K. F., and Roberts D. W., 1989. A Frameless stereotaxic operating microscope for neurosurgery. *IEEE Transactions on Biomedical Engineering*, vol. **36**, pp. 608–617.

[31] Roberts D. W., Friets E. M., Strohbehn J. W., and Nakajima T., 1993. The sonic digitizing microscope. In Maciunas R. J., Ed., *Interactive Image-Guided Neurosurgery, USA*, American Association of Neurological Surgeons.

[32] King A. P., Edwards P. J., Maurer Jr. C. R., de Cunha D. A., Gaston R. P., Clarkson M., Hill D. L. G., Hawkes D. J., Fenlon M. R., Strong A. J., Cox T. C. S., and Gleeson M. J., 2000. Stereo augmented reality in the surgical microscope, *Presence: Teleoperators and Virtual Environments*, vol. **9**, pp. 360–368.

[33] Sasama T., Sugano N., Sato Y., Momoi Y., Koyama T., Nakajima Y., Sakuma I., Fujie M. G., Yonenobu K., Ochi T., and Tamura S., 2002. A novel laser guidance system for alignment of linear surgical tools: its principles and performance evaluation as a man-machine system. In *5th International Conference on Medical Image Computing and Computer-Assisted Intervention*, pp. 125–132.

[34] Fischer G. S., Wamsley C., Zinreich S. J., and Fichtinger G., 2006. MRI guided needle insertion-comparison of four techniques. In *Annual Scientific Conference of the Society of Interventional Radiology*, Toronto, Canada.

[35] Uecker D. R., Lee C., Wang Y. F., and Wang Y., 1994. A speech-directed multi-modal man-machine interface for robotically enhanced surgery. In *First Int. Symp. on Medical Robotics and Computer Assisted Surgery (MRCAS 94)*, Pittsburgh, pp. 176–183.

[36] Gupta P. K., 2001. *A Method to Enhance Microsurgical Tactile Perception and Performance Through the Use of Auditory Sensory Perception*, Master of Science Thesis in M.S. in Engineering, The Johns Hopkins University, Baltimore.

[37] Abovitz R. A., and Quaid A. E., 2001. The future use of networked haptic learning information systems in computer-assisted surgery. In *Proc. CAOS USA 2001*, Pittsburgh, pp. 337–338.

[38] Gerovich O., Marayong P., and Okamura A. M., 2004. The effect of visual and haptic feedback on computer-assisted needle insertion. *Computer-Aided Surgery*, vol. **9**, pp. 243–249.

[39] Okamura A. M., 2004. Methods for haptic feedback in teleoperated robot-assisted surgery. *Industrial Robot*, vol. **31**, pp. 499–508.

[40] Quaid A. E., and Abovitz R. A., 2002. Haptic information dispays for computer-assisted surgery. In *IEEE International Conference on Robotics and Automation*, pp. 2092–2097.

[41] Akinbiyi T., Reiley C. E., Saha S., Burschka D., Hasser C. J., Yuh D. D., and Okamura A. M., 2006. Dynamic augmented reality for sensory substitution in robot-assisted surgical systems. In *28th Annual International Conference of the IEEE Engineering in Medicine and Biology Society*, pp. 567–570.

[42] Blackwell M., Nikou C., DiGioia A. M., and Kanade T., 2000. An image overlay system for medical data visualization. *Medical Image Analysis*, vol. **4**, pp. 67–72.

[43] Fichtinger G., Degeut A., M. K., Balogh E., Fischer G., Mathieu H., Taylor R. H., Zinreich S., and Fayad L. M., 2005. Image overlay guidance for needle insertion on CT scanner. *IEEE Trans on Biomedical Engineering*, vol. **52**, pp. 1415–1424.

[44] Fischer G. S., Deguet A., Fayad L. M., Zinreich S. J., Taylor R. H., and Fichtinger G., 2006. Musculoskeletal needle placement with MRI image overlay guidance. In *Annual Meeting of the International Society for Computer Assisted Surgery*, Montreal, Canada, pp. 158–160.

[45] Leven J., Burschka D., Kumar R., Zhang G., Blumenkranz S., Dai X., Awad M., Hager G. D., Marohn M., Choti M., Hasser C. J., and Taylor R. H., 2005. DaVinci canvas: a telerobotic surgical system with integrated, robot-assisted, laparoscopic ultrasound capability. In *MICCAI*, pp. 811–818.

[46] Mettler L., Ibrahim M., and Jonat W., October 1998. One year of experience working with the aid of a robotic assistant (the voice-controlled optic holder AESOP) in gynaecological endoscopic surgery., *Hum Reprod.*, vol. **13**, pp. 2748–2750.

[47] Reichenspurner H., Demaino R., Mack M., Boehm D., Gulbins H., Detter C., Meiser B., Ellgass R., and Reichart B., 1999. Use of the voice controlled and computer-assisted surgical system Zeus for endoscopic coronary artery surgery bypass grafting., *J. Thoracic and Cardiovascular Surgery*, vol. **118**.

[48] Sturges R., and Laowattana S., 1996. A voice-actuated, tendon-controlled device for endoscopy. In Taylor R. H., Lavallee S., Burdea G., and Mosges R., Eds. *Computer-Integrated Surgery*, Cambridge, Mass.: MIT Press.

[49] Confer R. G., and Bainbridge R. C., 1984. Voice control in the microsurgical suite. In *Proc. of the Voice I/O Systems Applications Conference '84*, Arlington, Va.,

[50] Nishikawa A., Hosoi T., Koara K., and Dohi T., October 2003. FAce MOUSE: a novel human-machine interface for controlling the position of a laparoscope. *IEEE Trans on Robotics and Automation*, vol. **19**, pp. 818–824.

[51] Taylor R. H., Funda J., Eldridge B., Gruben K., LaRose D., Gomory S., Talamini M., MD, Kavoussi L., MD, and Anderson J., 1995. A telerobotic assistant for laparoscopic surgery. In *IEEE EMBS Magazine Special Issue on Robotics in Surgery*, pp. 279–291.

[52] Krupa A., Gangloff J., Doignon C., deMathelin M. F., Morel G., Leroy J., Soler L., and Marescaux J., October 2003. Autonomous 3D positioning of surgical instruments in robotized laparoscopic surgery using visual servoing. *IEEE Trans on Robotics and Automation*, vol. **19**, pp. 842–853.

[53] Fischer G., Akinbiyi T., Saha S., Zand J., Talamini M., Marohn M., and Taylor, R. H., 2006. Ischemia and force sensing surgical instruments for augmenting available surgeon information. In *IEEE International Conference on Biomedical Robotics and Biomechatronics – BioRob 2006*, Pisa, Italy.

[54] Fischer G. S., Zand J., Marohn M., Akinbiyi T., Kanev K., Kuo J., Kazanzides P., and Taylor R. H., 2005. Intraoperative ischemia sensing surgical instruments. In *International Conference on Complex Medical Engineering*, Takamatsu, Japan.

[55] Fischer G., Saha S., Horwat J., Yu J., Zand J., Marohn M., Talamini M., and Taylor R. H., 2005. An intra-operative system for relating ischemic damage to retraction forces. In *BMES*, Baltimore, MD.

[56] Morimoto A., Foral R., Kuhlman J., Zucker K., Curet M., Bockalage T., MacFarlane T., and Kory L., 1997. Force sensor for laparoscopic Babcock. In *Medicine Meets Virtual Reality*, pp. 354–361.

[57] Bicchi A., Canepa G., Rossi D. D., Iacconi P., and Scilingo E., 1996. A sensorised minimally invasive surgery tool for detecting tissutal elastic properties. In *Proc. of the IEEE International Conference on Robotics and Automation*, pp. 884–888.

[58] Poulose B., Kutka M., Sagaon M. M., Barnes A., Yang C., Taylor R., and Talamini M., 1998. Human versus robotic organ retraction during laparoscopic Nissen Fundoplication. In

Medical Image Computing and Computer-Assisted Interventions (MICCAI-98), Cambridge, Mass. pp. 197–206.

[59] Poulose P. K., Kutka M. F., Mendoza-Sagaon M., Barnes A. C., Yang C., Taylor R. H., and Talamini M. A., 1999. Human vs robotic organ retraction during laparoscopic Nissen Fundoplication. *Surgical Endoscopy*, vol. **13**, pp. 461–465.

[60] Prasad S., Kitagawa M., Fischer G. S., Zand J., Talamini M. A., Taylor R. H., and Okamura A. M., 2003. A modular 2-DOF force-sensing instrument for laparoscopic surgery. In *Conference on Medical Image Computing and Computer Assisted Intervention*, Montreal, pp. 279–286.

[61] Gupta P., Jensen P, and de Juan E., 1999. Quantification of tactile sensation during retinal micro-surgery. In *MICCAI99: The Second International Conference on Medical Image Computing and Computer-Assisted Intervention*, Cambridge, England.

[62] Gupta P., Jensen P., and de Juan E., 1999. Surgical forces and tactile perception during retinal microsurgery. In *MICCAI*, pp. 1218–1225.

[63] Rosen J., Brown J. D., Chang L., Barreca M., Sinanan M., and Hannaford B., 2002. The Blue-DRAGON – a system for measuring the kinematics and the dynamics of minimally invasive surgical tools in-vivo. In *IEEE International Conference on Robotics and Automation*, pp. 1876–1881.

[64] Rosen J., Hannaford B., MacFarlane M., and Sinanan M., 1999. Force controlled and teleoperated endoscopic grasper for minimally invasive surgery – experimental performance evaluation. *IEEE Transactions on Biomedical Engineering*, vol. **46**, pp. 1212–1221.

[65] Menciassi A., Eisinberg A., Carrozza M. C., and Dario P., 2003. Force sensing microinstrument for measuring tissue properties and pulse in microsurgery. *IEEE/ASME Transactions on Mechatronics*, vol. **8**, pp. 10–17.

[66] Berkelman P. J., Whitcomb L., Taylor R., and Jensen P., October 2003. A miniature microsurgical instrument tip force sensor for enhanced force feedback during robot-assisted manipulation. *IEEE T. Robotics and Automation*, vol. **19**, pp. 917–922.

[67] Rothbaum D. L., Roy J., Berkelman P., Hager G., Stoianovici D., Taylor R. H., Whitcomb L. L., Howard Francis M., and Niparko J. K., November 2002. Robot-assisted stapedotomy: micropick fenestration of the stapes footplate, *Otolaryngology – Head and Neck Surgery*, vol. **127**, pp. 417–426.

[68] Howe R. D., Peine W. J., Kontarinis D. A., and Son J. S., 1995. Remote palpation technology. *IEEE Engineering in Medicine and Biology*, vol. **14**, pp. 318–323.

[69] Beasly R., and Howe R., 2002. Tactile tracking of arteries in robotic surgery. In *IEEE International Conference on Robotics and Automation*, pp. 3801–3806.

[70] Ottensmeyer M. P., and Salisbury J. K., 2001. In vivo data acquisition instrument for solid organ mechanical property measurement. *Proceedings of the Medical Image Computing and Computer-Assisted Intervention 4th International Conference*, pp. 975–982.

[71] Brouwer I., Ustin J., Bentley L., Sherman A., Dhruv N., and Tendick F., 2001. Measuring in vivo animal soft tissue properties for haptic modeling in surgical simulation. In *Medicine Meets Virtual Reality*, Westwood J. D., Ed., IOS Press, Amsterdam, pp. 69–74.

[72] Solomon S. B., Patriciu A., Bohlman M. E., Kavoussi L. R., and Stoianovici D., 2002. Robotically driven interventions: a method of using CT fluoroscopy without radiation exposure to the physician. *Radiology*, vol. **225**, pp. 277–282.

[73] Patriciu A., Solomon S., Kavoussi L. R., and Stoianovici D., 2001. Robotic kidney and spine percutaneous procedures using a new laser-based CT registration method. In *Proceedings to Medical Image Computing and Computer-Assisted Intervention*, pp. 249–257.

[74] Lozano-Pérez T., July 1983. Robot programming. *Proceedings of the IEEE*, vol. **71**.

[75] Kazanzides P., Deguet A., Kapoor A., Sadowsky O., LaMora A., and Taylor R., 2005. Development of open source software for computer-assisted intervention systems. In

ISC/NAMIC/MICCAI Workshop on Open-Source Software (also available online at Insight Journal, http://hdl.handle.net/1926/46), Palm Springs, CA.

[76] Kapoor A., Deguet A., and Kazanzides P., 2006. Software components and frameworks for medical robot control. In *Proc. IEEE Intl. Conf. on Robotics and Automation*, pp. 3813–3818.

[77] Kazanzides P., DiMaio S., Cleary K., Fichtinger G., and Taylor R., 2006. System architecture and toolkits for image-guided intervention systems. In *Medicine Meets Virtual Reality 14*, Long Beach, California.

[78] Maurer C., Fitzpatrick J., Wang M., Galloway R., Maciunas R., and Allen G., 1997. Registration of head volume images using implantable fiducial markers. *IEEE Trans. on Medical Imaging*, vol. **16**, pp. 447–462.

[79] Levensen N. G., and Turner C. S., 1993. An investigation of the Therac-25 accidents. *Computer*, vol. **26**, pp. 18–41.

[80] McDermott R. E., Mikulak R. J., and Beauregard M. R., 1996. *The Basics of FMEA: Quality Resources*.

[81] Pott P., Scharf H., and Schwarz M., March 2005. Today's state of the art in surgical robotics. *Computer Aided Surgery*, vol. **10**, pp. 101–132.

[82] Mittelstadt B., Kazanzides P., Zuhars J., Williamson B., Cain P., Smith F., and Bargar W., 1996. The evolution of a surgical robot from prototype to human clinical use. In Taylor R. H., Lavallee S., Burdea G., and Mosges R., Eds., *Computer-Integrated Surgery*, Cambridge, Mass.: MIT Press, pp. 397–407.

[83] Taylor R. H., Paul H. A., Kazanzides P., Mittelstadt B. D., Hanson W., Zuhars J. F., Williamson B., Musits B. L., Glassman E., and Bargar W. L., 1994. An image-directed robotic system for precise orthopaedic surgery. *IEEE Transactions on Robotics and Automation*, vol. **10**, pp. 261–275.

[84] Garbini J. L., Kaiura R. G., Sidles J. A., Larson R. V., and Matson F. A., 1987. Robotic instrumentation in total knee arthroplasty. In *Proc. 33rd Annual Meeting, Orthopaedic Research Society*, San Francisco, p. 413.

[85] Jakopec M., Harris S. J., Baena F. R. Y., Gomes P., Cobb J., and Davies B. L., 2001. The first clinical application of a hands-on robotic knee surgery system. *Computer Aided Surgery*, vol. **6**, pp. 329–339.

[86] Kwon D. S., Lee J. J., Yoon Y. S., Ko S. Y., Kim J., Chung J. H., Won C. H., and Kim J. H., 2002. The mechanism and the registration method of a surgical robot for hip arthroplasty. In *IEEE International Conference on Robotics and Automation*, pp. 1889–2949.

[87] Sugita N. M. N., Warisawa S., Mitsuishi M., Fujiwara K., Abe N., Inoue T., Kuramoto K., Nakashima Y., Tanimoto K., Suzuki M., Moriya H., and Hashizume H., 2006. Development of a computer-integrated minimally invasive surgical system for knee arthroplasty. In *IEEE/RAS-EMBS International Conference Biomedical Robotics and Biomechatronics*.

[88] Marcacci S., Dario P., Fadda M., Marcenaro G., and Martelli S., 1996. Computer-assisted knee arthroplasty. In Taylor R. H., Lavallee S., Burdea G., and Mosges R., Eds., *Computer-Integrated Surgery*, Cambridge, Mass.: MIT Press, pp. 417–423.

[89] Siebert W., and Mai S., 2001. One year clinical experience using the robot system CASPAR for TKR. In *Proc. CAOS USA 2001*, Pittsburgh, pp. 141–142.

[90] Bargar W., DiGioia A., Turner R., Taylor J., McCarthy J., and Mears D., 1995. Robodoc multicenter trial: an interim report. In *Proc. 2nd Int. Symp. on Medical Robotics and Computer Assisted Surgery*, Baltimore, Md., pp. 208–214.

[91] Skibbe H., Börner M., Wiesel U., and Lahmer A., 1999. Revision THR using the ROBODOC system. In *CAOS/USA '99*, Pittsburgh, Pennsylvania, USA, pp. 110–111.

[92] Wiesel U., Lahmer A., Tenbusch M., and Börner M., 2001. Total knee replacement using the Robodoc system. In *Proc. First Annual Meeting of CAOS International*, Davos, p. 88.

[93] Paul H., Bargar W., Mittelstadt B., Musits B., Taylor R., Kazanzides P., et al., December 1992. Development of a surgical robot for cementless total hip arthroplasty. *Clinical Orthopaedics and Related Research*, vol. **285**, pp. 57–66.

[94] Kazanzides P., Mittelstadt B. D., Musits B. L., Bargar W. L., Zuhars J. F., et al., 1995. An integrated system for cementless hip replacement. *IEEE Engineering in Medicine and Biology*, vol. **14**, pp. 307–313.

[95] Cohan S., 2001. ROBODOC achieves pinless registration. *Industrial Robot*, vol. **28**, pp. 381–386.

[96] Zuhars J., and Hsia T., 1995. Nonhomogeneous material milling using a robot manipulator with force controlled velocity. In *IEEE Intl. Conf. on Robotics and Automation*, Nagoya, Japan, pp. 1461–1467.

[97] Kwoh Y. S., Hou J., Jonckheere E. A., and Hayati S., 1988. A robot with improved absolute positioning accuracy for CT guided stereotactic brain surgery. *IEEE Transactions on Biomedical Engineering*, vol. **35**, pp. 153–160.

[98] Benabid A. L., Cinquin P., Lavalle S., Le Bas J. F., Demongeot J., and de Rougemont J., 1987. Computer-driven robot for stereotactic surgery connected to CT scan and magnetic resonance imaging. Technological design and preliminary results. *Appl. Neurophysiol*, vol. **50**, pp. 153–154.

[99] Masamune K., Kobayashi E., Masutani Y., Suzuki M., Dohi T., Iseki H., and Takakura K., 1995. Development of an MRI-compatible needle insertion manipulator for stereotactic neurosurgery. *Journal of Image Guided Surgery*, vol. **1**, pp. 242–248.

[100] Li Q., Zamorano L., Pandya A., Perez R., Gong J., and Diaz F., 2002. The application accuracy of the NeuroMate robot–A quantitative comparison with frameless and frame-based surgical localization systems. *Computer Assisted Surgery*, vol. **7**, pp. 90–98.

[101] Masamune K., Fichtinger G., Patriciu A., Susil R. C., Taylor R. H., Kavoussi L. R., Anderson J. H., Sakuma I., Dohi T., and Stoianovici D., 2001. System for robotically assisted percutaneous procedures with computed tomography guidance. *Journal of Computer Assisted Surgery*, vol. **6**, pp. 370–383.

[102] Jain A., Mustufa T., Zhou Y., Burdette E., Chirikjian G., and F. G, 2005. Robust fluoroscope tracking fiducial. *Med Phys.*, vol. **32**, pp. 3185–3198.

[103] Susil R. C., Ménard C., Krieger A., Coleman J. A., Camphausen K., Choyke P., Ullman K., Smith S., Fichtinger G., Whitcomb L. L., Coleman N., and Atalar E., January 2006. Transrectal prostate biopsy and fiducial marker placement in a standard 1.5T MRI scanner. *J. Urol.*, vol. **175**, pp. 113–120.

[104] Chinzei K., Kikinis R., and Jolesz F. A., 1999. MR compatibility of mechatronic devices: design criteria. In *Second International Conference on Medical Image Computing and Computer-Assisted Intervention*, pp. 1020–1030.

[105] Watanabe E., Watanabe T., and Manka S., et. al., 1987. Three-dimensional digitizer (neuronavigator): new equipment for computed tomography-guided stereotaxic surgery, *Surg. Neurol.*, vol. **27**, pp. 543–547.

[106] Adams L., Knepper A., Krybus W., Meyer-Ebrecht D., Pfeifer G., Ruger R., and Witte M., 1992. Orientation aid for head and neck surgeons. *Innovation et Technologie en Biologie et Medicine*, vol. **14**, pp. 409–424.

[107] VanderKolk C., Zinreich S., Carson B., Bryan N., and Manson P., 1992. An interactive 3D-CT surgical localizer for craniofacial surgery. In *Craniofacial Surgery*, Bologna, Italy, p. 25.

[108] Cutting C. B., Bookstein F. L., and Taylor R. H., 1996. Applications of simulation, morphometrics and robotics in craniofacial surgery. In Taylor R. H., Lavallee S., Burdea G., and Mosges R., Eds., *Computer-Integrated Surgery*, Cambridge, Mass.: MIT Press, pp. 641–662.

[109] Schenker P. S., Das H. O., and Timothy R., 1995. Development of a new high-dexterity manipulator for robot-assisted microsurgery. In *Proceedings of SPIE – The International Society for Optical Engineering: Telemanipulator and Telepresence Technologies*, Boston, MA, pp. 191–198.

[110] Cavusoglu M. C., Williams W., Tendick F., and Sastry S., January 2003. Robotics for Telesurgery: Second Generation Berkeley/UCSF Laparoscopic telesurgical workstation and looking towards the future applications. *Industrial Robot*, vol. **30**, pp. 22–29.

[111] Salcudean S. E., Ku S., and Bell G., 1997. Performance measurement in scaled teleoperation for microsurgery. In *Proc. First Joint Conference of CVRMed and MRCAS*, Grenoble, France, pp. 789–798.

[112] Ikuta K., Hasegawa T., and Daifu S., 2003. Hyper redundant miniature manipulator 'hyper finger' for remote minimally invasive surgery in deep area, in *IEEE Conference on Robotics and Automation*, Taiwan, pp. 1098–1102.

[113] Hongo K., Kobayashi S., Kakizawa Y., Koyama J.-I., Goto T., Okudera H., Kan K., Fujie M. G., Iseki H., and Takakura K., October 2002. NeuRobot: telecontrolled micromanipulator system for minimally invasive microneurosurgery – preliminary results. *Neurosurgery*, vol. **51**, pp. 985–988.

[114] Kanade T., Davies B., and Riviere C., 2006. Special issue on medical robotics. *IEEE Proceedings*, vol. **94**.

[115] Guthart G. S., and Salisbury J. K., 2000. The intuitive telesurgery system: overview and application. In *Proc. of the IEEE International Conference on Robotics and Automation (ICRA2000)*, San Francisco, pp. 618–621.

[116] Marescaux J., and Rubino F., 2003. The ZEUS robotic system: experimental and clinical applications, *Surg. Clin. North Amer.*, vol. **83**, pp. 1305–1315.

[117] Taylor R., Jensen P., Whitcomb L., Barnes A., Kumar R., Stoianovici D., Gupta P., Wang Z., deJuan E., and Kavoussi L., 1999. A steady-hand robotic system for microsurgical augmentation. *International Journal of Robotics Research*, vol. **18**.

[118] Gomez-Blanco M. A., Riviere C. N., and Khosla P. K., 2000. Intraoperative tremor monitoring for vitreoretinal microsurgery. In *Proc. Medicine Meets Virtual Reality* 8, pp. 99–101.

[119] Kapoor A., Kumar R., and Taylor R., 2003. Simple biomanipulation tasks with a 'steady hand' cooperative manipulator. In *Proceedings of the Sixth International Conference on Medical Image Computing and Computer Assisted Intervention – MICCAI 2003*, Montreal, pp. 141–148.

[120] Iordachita I., Kapoor A., Mitchell B., Kazanzides P., Hager G., Handa J., and Taylor R., 2006. Steady-hand manipulator for retinal surgery. In *MICCAI Workshop on Medical Robotics*, Copenhagen, pp. 66–73.

[121] Ikuta K., Yamamoto K., and Sasaki K., 2003. Development of remote microsurgery robot and new surgical procedure for deep and narrow space. In *IEEE Conference on Robotics and Automation*, Taiwan, pp. 1103–1108.

[122] Suzuki N., Hayashibe M., and Hattori A., 2005. Development of a downsized master-slave surgical robot system for intragastic surgery. In *ICRA Surgical Robotics Workshop*, Barcelona, Spain.

[123] Ikuta K., Ichikawa H., Suzuki K., and Yajima D., 2006. Multi-degree of freedom hydraulic pressure driven safety active catheter. In *Proceedings of the 2006 IEEE International Conference on Robotics and Automation*, Orlando, Florida, pp. 4161–4166.

[124] Ascari L., Stefanini C., Menciassi A., Sahoo S., Rabischong P., and Dario P., 2003. A new active microendoscope for exploring the sub-arachnoid space in the spinal cord. In *IEEE Conference on Robotics and Automation*, pp. 2657–2662.

[125] Grundfest S. W., Burdick J. W., and Slatkin A. B., 1995. The development of a robotic endoscope. In *IEEE International Conference on Robotics and Automation*, Nagoya, Japan, pp. 162–171.

[126] Stefanini C., Menciassi A., and Dario P., May-June 2006. Modeling and experiments on a legged microrobot locomoting in a tubular, compliant and slippery environment. *International Journal of Robotics Research*, vol. **25**, pp. 551–560.

[127] Patronik N., Riviere C., Qarra S. E., and Zenati M. A., 2005. The HeartLander: a novel epicardial crawling robot for myocardial injections. In *Proceedings of the 19th International Congress of Computer Assisted Radiology and Surgery*, pp. 735–739.

[128] Simaan N., Taylor R., and Flint P., 2004. High dexterity snake-like robotic slaves for minimally invasive telesurgery of the throat. In *Int. Symp. on Medical Image Computing and Computer-Assisted Interventions*, pp. 17–24.

[129] Simaan N., Taylor R., Hillel A., and Flint P., 2007. Minimally invasive surgery of the upper airways: addressing the challenges of dexterity enhancement in confined spaces. In R. Faust, Ed., *Robotics in Surgery – History, Current and Future Applications*, Nova Science Publishing, pp. 261–280.

[130] Goradia T. M., Taylor R. H., and Auer L. M., 1997. Robot-assisted minimally invasive neurosurgical procedures: first experimental experience. In *Proc. First Joint Conference of CVRMed and MRCAS*, Grenoble, France, pp. 319–322.

[131] Taylor R., Jensen P., Whitcomb L., Barnes A., Kumar R., Stoianovici D., Gupta P., Wang Z. X., deJuan E., and Kavoussi L., 1999. Steady-hand robotic system for microsurgical augmentation, *International Journal of Robotics Research*, vol. **18**, pp. 1201–1210.

[132] Riviere C. V., Rader R. S., and Thakor N. V., 1995. Adaptive real-time cancelling of physiological tremor for microsurgery. In *Proc. 2nd Int. Symp. on Medical Robotics and Computer Assisted Surgery (MRCAS)*, Baltimore, Md.

[133] Ang W. T., and Riviere C. N., 2001. Neural network methods for error canceling in human-machine manipulation. In *22nd Annu. Conf. IEEE Eng. Med. Biol. Soc*, Istanbul, pp. 3462–3465.

[134] Xu S., Fichtinger G., Taylor R. H., and Cleary K., 2004. 3D motion tracking of pulmonary lesions using CT fluoroscopy images for robotically assisted lung biopsy. In *SPIE International Society of Optical Engineering*, pp. 394–402.

[135] Shoham M., Burman M., Zehavi E., Joskowicz L., Batkilin E., and Kunicher Y., 2003. Bone-mounted miniature robot for surgical procedures: concept and clinical applications. *IEEE Transactions on Robotics and Automation*, vol. **19**, pp. 893–901.

[136] Plaskos C., Cinquin P., Lavallee S., and Hodgson A. J., December 2005. Praxiteles: a miniature bone-mounted robot for minimal access total knee arthroplasty, *Int. J. of Medical Robotics and Computer Assisted Surgery*, vol. **1**, pp. 67–79.

[137] McEwen J. A., Bussani C. R., Auchinleck G. F., and Breault M. J., 1989. Development and initial clinical evaluation of pre-robotic and robotic retraction systems for surgery. In *Proc. Second Workshop on Medical and Health Care Robotics*, Newcastle-onTyne, pp. 91–101.

[138] Kochan A., 2005. Scalpel please, robot: Penelope's debut in the operating room. *Industrial Robot*, vol. **32**, pp. 449–451.

[139] Miyawaki F., Masamune K., Suzuki S., Yoshimitsu K., and Vain J., 2005. Scrub nurse robot system-intraoperative motion analysis of a scrub nurse and timed-automata-based model for surgery. *IEEE Transactions on Industrial Electronics*, vol. **52**, pp. 1227–1235.

[140] Begin E., Gagner M., and Hurteau R., 1995. A robotic camera for laparoscopic surgery: conception and experimental results. *Surgical Laparoscopy & Endoscopy*, vol. **5**.

[141] Sackier J. M., and Wang Y., January 1994. Robotically assisted laparoscopic surgery, from concept to development. *Surg Endosc*, vol. **8**, pp. 63–66.

[142] Satava R., 1992. Robotics, telepresence, and virtual reality: a critical analysis of the future of surgery. *Minimally Invasive Therapy*, vol. **1**, pp. 357–363.

[143] Satava R., 1995. Virtual reality, telesurgery, and the new world order of medicine. *Journal of Image-Guided Surgery*, vol. **1**, pp. 12–16.
[144] Green P., Satava R., Hill J., and Simon I., 1992. Telepresence: advanced teleoperator technology of minimally invasive surgery (abstract), *Surgical Endoscopy*, vol. **6**.
[145] Mitsuishi M., Warisawa S. I., Tsuda T., Higuchi T., Koizumi N., Hashizume H., and Fujiwara K., 2001. Remote ultrasound diagnostic system. In *Proc. IEEE Conf. on Robotics and Automation*, Seoul, pp. 1567–1574.
[146] Cunha D. D., Gravez P., Leroy C., Maillard E., Jouan J., Varley P., Jones M., Halliwell M., Hawkes D., Wells P. N. T., and Angelini L., 1998. The MIDSTEP system for ultrasound guided remote telesurgery, In *IEEE EMBS*, pp. 1266–1269.
[147] Lee B. R., Stoianovici D., Bishoff J. T., Micali S., Micali F., Bauer J., Whitcomb L. L., Taylor R. H., and Kavoussi L. R., 1999. TELEPAKY: a new robotic system for active remote telesurgery. *The Lancet.*
[148] Bauer J., Lee B. R., Stoianovici D., Bishoff J. T., Micali S., Micali F., and Kavoussi L. R., Winter 2001. Remote percutaneous renal access using a new automated telesurgical robotic system. *Telemed J E Health*, vol. **7**, pp. 341–346.
[149] Frimberger D., Kavoussi L. R., Stoianovici D., Adam C., Zaak D., Corvin S., Hofstetter A., and Oberneder R., 2002. Telerobotische Chirurgie zwischen Baltimore und München, *Der Urologe [A]*, vol. **41**, pp. 489–492.
[150] Marescaux J., Leroy J., Gagner M., Rubino F., Mutter D., Vix M., Butner S. E., and Smith M. K., September 27 2001. Transatlantic robot-assisted telesurgery. *Nature*, vol. **413**, pp. 379–380.
[151] Anvari M., Broderick T., Stein H., Chapman T., Ghodoussi M., Birch D. W., Mckinley C., Trudeau P., Dutta S., and Goldsmith C. H., March 2005. The impact of latency on surgical precision and task completion during robotic-assisted remote telepresence surgery. *Comput Aided Surg*, vol. **10**, pp. 93–99.
[152] Dotto L., 2006. Application – revolutionary telemedicine techniques, Summary online article about Anvari's remote telesurgery work.
[153] Kavoussi L., Moore R., Partin A., Bender J., Venilman M., and Satava R., 1994. Telerobotic-assisted laparoscopic surgery: initial laboratory and clinical experience. *Urology*, vol. **44**, pp. 15–19.
[154] Hanly E., Miller B., Kumar R., Hasser C., Coste-Maniere E., Talamini M., Aurora A., Schenkman N., and Marohn M., 2006. Mentoring console improves collaboration and teaching in surgical robotics. *J. Laparoendosc Adv. Surg. Tech.*, vol. **16**, pp. 445–451.
[155] Herman B., Hanly E., Schenkman N., Taylor R., Talamini M., and Marohn M., 2005. Telerobotic surgery creates opportunity for augmented reality surgery. *Telemedicine Journal and e-Health*, vol. **11**, p. 203.
[156] Rosenberg L. B., 1993. Virtual Fixtures: Perceptual tools for telerobotic manipulation. *Proceedings of IEEE Virtual Reality International Symposium*, pp. 76–82.
[157] Park S., Howe R. D., and Torchiana D. F., 2001. Virtual fixtures for robotic cardiac surgery, *Fourth International Conference on Medical Image Computing and Computer-Assisted Intervention.*
[158] Li M., and Okamura A. M., 2003. Recognition of operator motions for real-time assistance using virtual fixtures. In *11th International Symposium on Haptic Interfaces for Virtual Environment and Teleoperator Systems*, pp. 125–131.
[159] Marayong P., Bettini A., and Okamura A., 2002. Effect of Virtual Fixture Compliance on Human-Machine Cooperative Manipulation. In *IEEE/RSJ International Conference on Intelligent Robots and Systems*, pp. 1089–1095.
[160] Li M. and Taylor R. H., 2004. Spatial Motion Constraints in Medical Robots Using Virtual Fixtures Generated by Anatomy. In *IEEE Conf. on Robotics and Automation*, New Orleans, pp. 1270–1275.

[161] Lin H. C., Shafran I., Murphy T. E., Okamura A. M., Yuh D. D., and Hager G. D., 2005. Automatic detection and segmentation of robot-assisted surgical motions. In *MICCAI*, pp. 802–810.

[162] Mayer H., Nagy I., and Knoll A., 2003. Skill transfer and learning by demonstration in a realistic scenario of laparoscopic surgery. In *IEEE International Conference on Humanoids*, Munich, Germany.

[163] Kragic D., Marayong P., Li M., Okamura A. M., and Hager G. D., 2003. Human-machine collaborative systems for microsurgical applications. In *International Symposium on Robotics Research*.

[164] Li M., Ishii M., and Taylor R. H., 2007. Spatial motion constraints in medical robot using virtual fixtures generated by anatomy. *IEEE Transactions on Robotics*, vol. **23**, pp. 4–19.

[165] Kapoor A., Li M., and Taylor R. H., 2006. Constrained control for surgical assistant robots. In *IEEE Int. Conference on Robotics and Automation*, Orlando, pp. 231–236.

[166] Taylor R. H., Lavallee S., Burdea G. C., and Mosges R., 1996. Computer integrated surgery, MIT Press.

[167] Faust R., 2007. *Robotics in Surgery – History, Current and Future Applications*: Nova Science Publishing.

[168] Taylor R. H., Troccaz J., and Dario P., 2003. Special issue on medical robotics. *IEEE Transactions on Robotics and Automation*, vol. **19**.

Author Index

Numbers in *italics* indicate the pages on which complete references are given.

A

Abdel-Hamid, T.K., 20, *49*
Abdurazik, A., 63, 75, 78, *95, 96*
Abe, N., *96*, 237, *255*
Abovitz, R.A., 229, *252*
Acquisti, A., 210, *215*
Adam, C., 248, *259*
Adams, L., 243, *256*
Adrion, W.R., 20, *49*
Agarwal, R., 21, *50*
Akella, P., *251*
Akinbiyi, T., 229, 231, 247, *252, 253*
Akiyama, F., 27, *52*
Alavi, M., 21, *50*
Albin, J.L., 27, *52*
Albrecht, A.J., 24, *51*
Allen, G., 235, *255*
Alpert, S.R., 18, *48*
Ambler, S.W., 43, *55*
Amey, W.W., 20, *49*
Anderson, A., 23, 35, *51*
Anderson, J.H., 227, 229, 241, 248, *251, 253, 256*
Anderson, R.M., 16, *47*
Andrews, A., 58–62, 84, 85, 88, 94, 95, *95–97*
Ang, W.T., 247, *258*
Angelini, L., 248, *259*
Anvari, M., 248, *259*
Arango, G., 22, *51*
Arthur, L.J., 26, *52*
Ascari, L., 245, *257*
Atalar, E., 227, 242, *251, 256*

B

Auchinleck, G.F., 248, *258*
Auer, L.M., 247, *258*
Aurora, A., 248, *259*
Awad, M., 230, *253*
Azuma, M., 26, *52*

Bachman, C.W., 13, *46*
Badrinath, B.R., 115, 118, *158*
Baena, F.R.Y., 237, 239, 247, *255*
Bailey, J.E., 29, *54*
Bainbridge, R.C., 229, *253*
Baker, F.T., 19, *49*
Balcer, M.J., 84, *96*
Balogh, E., 230, *253*
Balsamo, S., 94, *95*
Banker, R.D., 20, *49*
Bargar, W., 237–240, 247, 248, *255, 256*
Barnes, A., 230, 247, *253, 254, 257, 258*
Barnes, S.J., 30, *54, 55*
Baroudi, J.J., *54*
Barreca, M., 230, *254*
Basili, V.R., 17, 24, 27, *48, 51*
Basin, D., 94, *95, 96*
Batkilin, E., 248, *258*
Bauer, F.L., 18, *48*
Bauer, J., 248, *259*
Beasly, R., 230, *254*
Beattie, R., 23, 35, *51*
Beauregard, M.R., 236, *255*
Bechtolsheim, A., 17, *47*
Beck, K., 22, 35, *51*
Begin, E., 248, *258*

261

Subject Index

Contents of Volumes in This Series

Printed and bound by CPI Group (UK) Ltd, Croydon, CR0 4YY

03/10/2024

01040416-0004